Lecture Notes in Mathematics

Edited by A. Dold and B. Eckmann

T0220229

1232

Peter Cornelis Schuur

Asymptotic Analysis of Soliton Problems

An Inverse Scattering Approach

Springer-Verlag

Berlin Heidelberg New York London Paris Tokyo

Author

Peter Cornelis Schuur
Mathematics Department, University of Technology
Den Dolech 2, 5600 MB Eindhoven, The Netherlands

Mathematics Subject Classification (1980): 15A60, 35B40, 35B45, 35P25, 35O20, 45M99, 47A99

ISBN 3-540-17203-3 Springer-Verlag Berlin Heidelberg New York
ISBN 0-387-17203-3 Springer-Verlag New York Berlin Heidelberg

This work is subject to copyright. All rights are reserved, whether the whole or part of the material is concerned, specifically those of translation, reprinting, re-use of illustrations, broadcasting, reproduction by photocopying machine or similar means, and storage in data banks. Under § 54 of the German Copyright Law where copies are made for other than private use, a fee is payable to "Verwertungsgesellschaft Wort", Munich.

© Springer-Verlag Berlin Heidelberg 1986
Printed in Germany

Printing and binding: Druckhaus Beltz, Hemsbach/Bergstr.
2146/3140-543210

How, in frames at rest,
the tail goes west,
while the east is won
by the soliton

PREFACE

A few years ago, when I started reading about solitons, I was fascinated by the beauty of the theory but at the same time astonished by the amount of assertions without even a shadow of a proof. No doubt, this lack of rigour is connected with the fact that many great discoveries were made during a relatively short period of time. In particular, in the early seventies it went more or less like this: A discovery was done and the proof was sketched. Immediately another discovery followed and the process repeated itself.

Of course, in this way a lot of questions remained unanswered. To mention only two of them: (i) How do KdV solitons emerge from arbitrary initial conditions? (ii) What are the phase shifts of these solitons as they interact both with the other solitons and with the dispersive wave-train?

The purpose of this volume is to provide answers to these and similar questions. Specifically, we give a complete, rigorous and explicit description of the emergence of solitons from various classes of nonlinear partial differential equations solvable by the inverse scattering technique. To this end we present an almost uniform method to obtain the asymptotic behaviour for large time of solutions of soliton problems in those coordinate regions where the nonsoliton component can be considered as a perturbation of the soliton component. The conditions under which our method works are remarkably general. For instance in the KdV analysis of Chapter 2 a mild algebraic decay of the initial function, so as to ensure that the associated reflection coefficient has a second derivative decaying at infinity as the inverse of its argument, is already sufficient.

The chapters in this volume are essentially self-contained with the exception of Chapter 1, which uses some concepts that are discussed in more detail in Chapter 2. Therefore the reader not particularly familiar with soliton theory is advised to read Chapter 2 before Chapter 1.

It is an honour to focus my thanks on two outstanding mathematicians, Wiktor Eckhaus and Aart van Harten, for their continuous interest in my work and for their pleasant way of combining moral support with valuable criticism.

To Wilma van Nieuwamerongen I am much indebted for skilfully typing the manuscript.

August 1986 Peter Schuur

TABLE OF CONTENTS

INTRODUCTION

For centuries nonlinearity formed a dark mystery. Nowadays, though things still look rather black, there are a few bright spots where we may confidently expect steady progress. This volume deals with one of these sparkles of hope: *the inverse scattering transformation*.

1. Historical remarks.

Many physical phenomena are nonlinear in nature. More often than not they can be modelled by nonlinear partial differential equations offering a wide range of complexity. Until the late sixties of this century the analyst had, roughly speaking, the choice: approximate or apologize. In the past two decades this situation changed, since various powerful nonperturbative mathematical techniques made their entrance. One of these is the inverse scattering technique (IST), also called inverse scattering transformation or spectral transform.

Its discovery is due to Gardner, Greene, Kruskal and Miura (GGKM for short) and was first reported in 1967 in their famous two-paged signal paper [9]. In this paper GGKM showed how to obtain the solution $u(x,t)$ of the Korteweg-de Vries (KdV) initial value problem

(1.1a) $u_t - 6uu_x + u_{xxx} = 0, \quad -\infty < x < +\infty, \quad t > 0$

(1.1b) $u(x,0) = u_0(x).$

Here and in the sequel a subscript variable indicates partial differentiation, e.g. $u_x = \frac{\partial u}{\partial x}$. Equation (1.1a) was first derived by Korteweg and de Vries [13] in 1895 in the context of free-surface gravity waves propagating in shallow water (see [4] for its historical background).

Below we shall discuss the GGKM method in some detail. Here we only mention its amazing starting point, namely the introduction of the solution u(x,t) of (1.1) as a potential in the Schrödinger scattering problem.

In 1968 Lax [16] put the GGKM method into a framework that clearly indicated its generality and had a substantial influence on future developments. In particular, Lax showed that (1.1a) is a member of an infinite family of nonlinear partial differential equations that can all be analysed in a similar fashion.

Guided by Lax' generalization of the pioneering work of GGKM, Zakharov and Shabat [26] were able to solve the initial value problem for another nonlinear equation of physical importance, the nonlinear Schrödinger equation (NLS)

(1.2) $iu_t = u_{xx} + 2|u|^2u.$

To this end they associated (1.2) with a spectral problem based on a system of two coupled first order ordinary differential equations. Incidentally, the NLS shows up in the description of plasma waves and models plane self-focusing and one-dimensional self-modulation.

Subsequently, Tanaka [20], [22] extended and rigorized the direct and inverse scattering theory for the Zakharov-Shabat system, motivated by the surprising discovery of Wadati [23] that another interesting nonlinear evolution equation could be solved by this system, namely the modified Korteweg-de Vries equation (mKdV)

(1.3) $u_t + 6u^2u_x + u_{xxx} = 0$

which appears in the continuum limit of a one-dimensional lattice with quartic anharmonicity [5].

Ablowitz, Kaup, Newell and Segur [1], [2] then showed that NLS and mKdV belong to a large class of nonlinear partial differential equations that can be solved via a generalized version of the Zakharov-Shabat scattering problem. Among these newly found integrable equations were several of physical importance, such as the sine-Gordon equation

$$(1.4) \qquad u_t = \tfrac{1}{2} \sin\left[2 \int_{-\infty}^{x} u(x',t)dx'\right]$$

which arises as an equation for the electric field in quantum optics [15], though the related forms

$$(1.4)' \qquad \sigma_{xt} = \sin \sigma \quad \text{and}$$

$$(1.4)'' \qquad \sigma_{xx} - \sigma_{tt} = \sin \sigma$$

appear more frequently in the literature (cf. [12]).

Herewith the triumphal march of the inverse scattering technique began. We shall not follow it further but refer to the survey articles [5], [10], [15], [17], [18] as well as the many textbooks on solitons [3], [6], [7], [8], [14], [25] currently available. We only mention that several other classes of physically relevant equations were found to be solvable by inverse scattering methods. In fact the process of finding new integrable nonlinear evolution equations has continued until this very day and has grown out into a major industry. Moreover, IST had its spin-off's to other areas of mathematics, like algebraic and differential geometry, functional and numerical analysis, etc. Nowadays - as stated in [6] - its applications range from nonlinear optics to hydrodynamics, from plasma to elementary particle physics, from lattice dynamics to electrical networks, from superconductivity to cosmology and geophysics. Moreover, IST is developing into an interdisciplinary subject, since it has recently penetrated in epidemiology and neurodynamics.

An essential reason for this wide applicability has not been mentioned so far: a dominant feature of nonlinear evolution equations of physical importance solvable via IST is that they admit exact solutions that describe the propagation and interaction of *solitons*.

At the moment there is no generally accepted mathematical definition of a soliton. As a working definition of a soliton we might take (cf. [5]) that it is a "localized" wave (in the sense of sufficiently rapidly decaying) which asymptotically preserves its shape and speed upon interaction with any other such localized wave. However, the concept of a soliton has a great intuitive appeal and is a good illustration of the fact that a happily chosen terminology is half of the success of a theory. The soliton was discovered in 1965 by Zabusky and Kruskal [24] while performing a numerical study of the KdV. Actually, the name "soliton" was suggested by Zabusky, who originally used the term "solitron" instead (see [6], pp. 176, 177).

Let us discuss their discovery in some detail. Already Korteweg and de Vries theirselves knew [13] that the KdV had a special travelling wave solution, the solitary wave

$$(1.5) \qquad u(x,t) = -2k_0^2 \operatorname{sech}^2[k_0(x - x_0 - 4k_0^2 t)], \qquad (\operatorname{sech} z = \frac{2}{e^z + e^{-z}})$$

where k_0 and x_0 are constants. Observe, that the velocity of this wave, $4k_0^2$, is proportional to its amplitude, $2k_0^2$. Now, in [24] Zabusky and Kruskal considered two waves such as (1.5), with the smallest to the right, as initial condition to the KdV. They discovered that after a certain time the waves overlap (the bigger one catches up), but that next the bigger one separates from the smaller and gradually the waves regain their initial shape and speed. The only permanent effect of the interaction is a phase shift, i.e. the center of each wave is at a different position than where it would have been if it had been travelling alone. Specifically, the bigger one is shifted to the right, the smaller to the left. The name soliton was chosen so as to stress this remarkable particle-like behaviour.

To conclude these introductory remarks, let us not forget to mention that, although in the past few years soliton interaction has been observed in various physical systems (see [3]), the first physical observation of what is now known as the single soliton solution (1.5) of the KdV already took place in the month of August 1834 by John Scott Russell, during his celebrated chase on horseback of a huge wave in the Union Canal, which from Edinburgh, joins with the Forth-Clyde canal and thence to the two coasts of Scotland. His own report of this experience, though classical by now, cannot be missed in any true soliton story.

It reads as follows [19]:

"I was observing the motion of a boat which was rapidly drawn along a narrow channel by a pair of horses, when the boat suddenly stopped - not so the mass of water in the channel which it had put in motion; it accumulated round the prow of the vessel in a state of violent agitation, then suddenly leaving it behind, rolled forward with great velocity, assuming the form of a large solitary elevation, a rounded, smooth and well defined heap of water, which continued its course along the channel apparently without change of form or diminution of speed. I followed it on horseback, and overtook it still rolling on at a rate of some eight or nine miles an hour, preserving its original figure some thirty feet long and a foot to a foot and a half in height. Its height gradually diminished, and after a chase of one or two miles I lost it in the windings of the channel. Such, in the month of August 1834, was my first chance interview with that singular and beautiful phenomenon ...".

2. IST for KdV: the gist of the method.

To comfort the reader who is completely new to the subject, let us at least give a rough sketch of how IST works, referring to [8] for the many intricate mathematical details. To this end we indicate here very briefly the basic features of the GGKM method, which is the first and undoubtedly the most fundamental example of an inverse scattering method.

Let us consider the KdV initial value problem (1.1) with $u_0(x)$ an arbitrary real function, sufficiently smooth and rapidly decaying for $x \to \pm\infty$. The surprising discovery of GGKM is now, that the nonlinear problem (1.1) can be solved in a series of linear steps, schematically representable in the following diagram

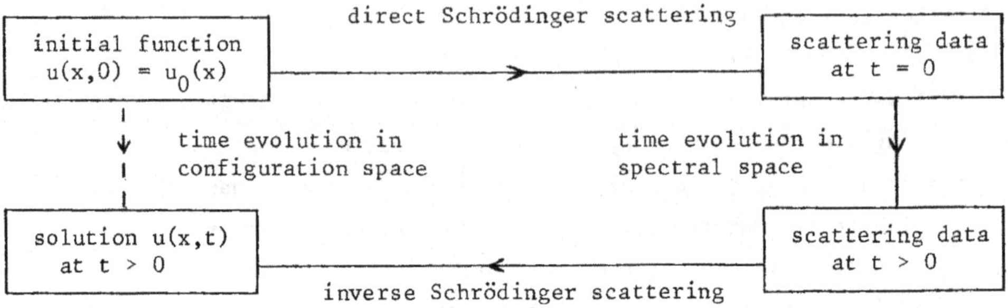

The manipulations suggested by this diagram are the following: For each $t \geqq 0$, introduce the real function $u(x,t)$ as a potential in the Schrödinger scattering problem

$$\psi_{xx} + (k^2 - u(x,t))\psi = 0, \qquad -\infty < x < +\infty. \tag{2.1}$$

For $t = 0$, compute the associated bound states $-\kappa_1^2 < -\kappa_2^2 < \ldots < -\kappa_N^2$, $\kappa_j > 0$, right normalization coefficients c_j^r and right reflection coefficient $b_r(k)$ (see Chapter 2 for their definition and properties), in other words, compute the right scattering data $\{b_r(k),\kappa_j,c_j^r\}$ associated with $u_0(x)$. Then, as $u(x,t)$ evolves according to the KdV, its right scattering data evolve in the following simple way:

$$\kappa_j(t) = \kappa_j \tag{2.2a}$$

$$c_j^r(t) = c_j^r \exp\{4\kappa_j^3 t\}, \qquad j = 1,2,\ldots,N \tag{2.2b}$$

$$b_r(k,t) = b_r(k)\exp\{8ik^3 t\}, \qquad -\infty < k < +\infty. \tag{2.2c}$$

To recover $u(x,t)$ from these data, one applies the inverse scattering procedure for the Schrödinger equation found by Gel'fand and Levitan [11], and defines

$$\Omega(\xi;t) = 2\sum_{j=1}^{N} [c_j^r(t)]^2 e^{-2\kappa_j\xi} + \frac{1}{\pi}\int_{-\infty}^{\infty} b_r(k,t)e^{2ik\xi}dk. \tag{2.3}$$

Next, one solves the Gel'fand-Levitan equation

$$\beta(y;x,t) + \Omega(x+y;t) + \int_0^{\infty} \Omega(x+y+z;t)\beta(z;x,t)dz = 0 \tag{2.4}$$

with $y > 0$, $x \in \mathbb{R}$, $t > 0$. The solution $\beta(y;x,t)$ has the important property

$$(2.5) \qquad \beta(0^+;x,t) = \int_x^\infty u(x',t)dx', \qquad x \in \mathbb{R}, \qquad t > 0,$$

and so we find that the solution of the KdV problem (1.1) is given by

$$(2.6) \qquad u(x,t) = -\frac{\partial}{\partial x}\beta(0^+;x,t), \qquad x \in \mathbb{R}, \qquad t > 0.$$

Notice that the original problem for the nonlinear partial differential equation (1.1) is essentially reduced in this way to the problem of solving a one-dimensional linear integral equation.

Explicit solutions of (2.4) have only been obtained for $b_r \equiv 0$. The solution $u_d(x,t)$ of the KdV with scattering data $\{0,\kappa_j,c_j^r(t)\}$ is called the pure N-soliton solution associated with $u_0(x)$, on account of its asymptotic behaviour displayed in the following remarkable result due to Tanaka [21]

$$(2.7a) \qquad \limsup_{t\to\infty}\ \sup_{x\in\mathbb{R}} \left|u_d(x,t) - \sum_{p=1}^N \left(-2\kappa_p^2 \operatorname{sech}^2[\kappa_p(x-x_p^+-4\kappa_p^2t)]\right)\right| = 0$$

$$(2.7b) \qquad x_p^+ = \frac{1}{2\kappa_p}\log\left\{\frac{[c_p^r]^2}{2\kappa_p}\prod_{\ell=1}^{p-1}\left(\frac{\kappa_\ell-\kappa_p}{\kappa_\ell+\kappa_p}\right)^2\right\}.$$

Thus for large positive time $u_d(x,t)$ arranges itself into a parade of N solitons with the largest one in front and this happens uniformly with respect to x on $\mathbb{R}$.

3. Asymptotics for nonzero reflection coefficient: main purpose of the book.

As illustrated by the previous section, the inverse scattering method enables us to obtain rather explicit exact solutions to nonlinear wave equations and to determine their asymptotic behaviour, which generally corresponds to a decomposition into solitons. Evidently, the problem of the asymptotic behaviour evolving from an arbitrary initial condition is in this way far from exhausted. It is still necessary to determine the asymptotic properties of the "nonsoliton part" of the solution whose presence is connected with the reflection coefficient being nonzero. In this volume we concern ourselves with this problem.

Rather than to give an elaborate general discussion, let us illustrate the ideas involved by considering again the KdV problem (1.1).

Suppose $u_0(x)$ is not a reflectionless potential. Then, in view of the fact that the linearized version of (1.1a) is a dispersive equation with associated group velocity $v_g = -3k^2 \leqq 0$, one expects that for large time the soliton part and the dispersive component will separate out, the dispersive wavetrain moving leftward and the solitons nicely arranging theirselves into a parade moving to the right similar to that described by (2.7). However, this is only heuristic reasoning. In fact it is dangerous reasoning too, since for nonlinear equations there is no such thing as a superposition principle.

The circumstance that at the time the question of validity of the above "plausible" conjecture had not been answered in a mathematically satisfactory way, formed the impetus for the research laid down in the present volume .

The main purpose of this book is therefore to give a complete and rigorous description of the emergence of solitons from various (classes of) nonlinear partial differential equations solvable by the inverse scattering technique.

Throughout the book we focus our attention on coordinate regions where the dispersive component is sufficiently small, e.g. $x \geqq -t^{1/3}$ for the (m)KdV problem. The behaviour of the solution in other regions, where the dispersive waves interact, is not discussed, since entirely different techniques are needed. For recent results in those regions we refer to [3].

4. Brief description of the contents.

The chapters in this volume are largely self-explanatory. Only Chapter 1 forms an exception. We therefore advise the reader new to the field to start with Chapter 2. In fact, both chapters deal with the KdV. However, in Chapter 1 the central ideas of our asymptotic method are exposed in the simplest nontrivial setting, whereas Chapter 2 serves to extend the results of Chapter 1, as well as to supply the details of the inverse scattering machinery. Also, the discussion of existence and uniqueness for the KdV initial value problem is postponed to Chapter 2.

In Chapter 1 we present a rigorous demonstration of the emergence of solitons from the KdV initial value problem with arbitrary real initial function. We show that for any choice of the constants $v > 0$ and $M \geq 0$ there exists a function $\sigma(t)$ tending to zero as $t \to \infty$, such that

$$(4.1) \qquad \sup_{x \geq -M+vt} |u(x,t) - u_d(x,t)| = O(\sigma(t)), \qquad \text{as } t \to \infty.$$

The exact behaviour of $\sigma(t)$ depends on properties of u_0. If u_0 decays exponentially for $x \to \pm\infty$, then so does $\sigma(t)$ for $t \to \infty$. If the decay of u_0 is only algebraic then also the decay of $\sigma(t)$ is algebraic.

In Chapter 2 we extend the asymptotic analysis given in Chapter 1. In fact, we no longer restrict our investigation to right half lines linearly moving rightward, but allow the right half lines to move slowly leftward. It is shown that in the absence of solitons the solution of (1.1) satisfies

$$(4.2) \qquad \sup_{x \geq -t^{1/3}} |u(x,t)| = O(t^{-2/3}), \qquad \text{as } t \to \infty,$$

whereas in the general case

$$(4.3) \qquad \sup_{x \geq -t^{1/3}} |u(x,t) - u_d(x,t)| = O(t^{-1/3}), \qquad \text{as } t \to \infty.$$

The emergence of solitons is clearly displayed by the remarkable convergence result

$$(4.4) \qquad \lim_{t\to\infty} \sup_{x \geq -t^{1/3}} \left| u(x,t) - \sum_{p=1}^{N} \left(-2\kappa_p^2 \operatorname{sech}^2[\kappa_p(x - x_p^+ - 4\kappa_p^2 t)] \right) \right| = 0$$

with x_p^+ as in (2.7b). In addition, we construct explicit x and t dependent bounds for the nonsoliton component of the solution and establish some interesting momentum and energy decomposition formulae. To support the analysis we only need to require - apart from the obvious assumption that IST works at all - that the right reflection coefficient b_r is of class $C^2(\mathbb{R})$ such that the derivatives $b_r^{(j)}(k)$, $j = 0,1,2$ behave as $O(|k|^{-1})$ for $k \to \pm\infty$. This condition is extremely weak. Hence our results apply to a large class of KdV initial value problems.

In Chapter 3 we study multisoliton solutions of the KdV in the general case of a nonzero reflection coefficient. We derive a new phase shift formula which shows that each soliton experiences in addition to the ordinary N-soliton phase shift an extra phase shift to the left caused by the interaction with the dispersive wavetrain. Moreover, we explicitly calculate this extra phase shift for the important case of a sech^2 initial function.

In Chapter 4 we consider the question how well a solution of a nonlinear wave equation is approximated by its soliton part in a more general setting: we derive an estimate which indicates how well a real potential in the Zakharov-Shabat system is approximated by its reflectionless part. Before doing so, the associated inverse scattering formalism is simplified considerably.

In Chapter 5 we use the estimate derived in Chapter 4 to obtain asymptotic bounds of solutions of the mKdV of the same type as those found in Chapter 2 for the KdV.

Using the results from Chapter 5 we derive in Chapter 6 a general phase shift formula for the mKdV remarkably similar in form to that found in the KdV case in Chapter 3. The only difference is, however, that now the extra phase shift is to the right, i.e. mKdV solitons are advanced by their interaction with the dispersive wavetrain.

Chapter 7 is devoted to the asymptotic analysis of the sine-Gordon equation on right half lines almost linearly moving leftward. Again the estimate found in Chapter 4 is shown to be of vital importance.

In Chapter 8 we study the Zakharov-Shabat system with complex potential and show that the results obtained in Chapter 4 can be generalized to this case. As an illustration we investigate the long-time behaviour of the solution of the complex modified Korteweg-de Vries initial value problem.

In an appendix we illustrate that our methods fail to give any result if the associated group velocity is not of constant sign. As an example we discuss the NLS and pose an interesting open problem.

References

[1] M.J. Ablowitz, D.J. Kaup, A.C. Newell and H. Segur, Method for solving the sine-Gordon equation, Phys. Rev. Lett. 30 (1973), 1262-1264.

[2] M.J. Ablowitz, D.J. Kaup, A.C. Newell and H. Segur, The inverse scattering transform - Fourier analysis for nonlinear problems, Stud. Appl. Math. 53 (1974), 249-315.

[3] M.J. Ablowitz and H. Segur, Solitons and the Inverse Scattering Transform, Philadelphia, SIAM, 1981.

[4] F. van der Blij, Some details of the history of the Korteweg-de Vries equation, Nieuw Archief voor Wiskunde 26 (1978), 54-64.

[5] R.K. Bullough and P.J. Caudrey, The soliton and its history, in: Solitons, Topics in Current Physics 17, Springer, New York, 1980 (edited by the same).

[6] F. Calogero and A. Degasperis, Spectral Transform and Solitons, Amsterdam, North-Holland, 1982.

[7] R.K. Dodd, J.C. Eilbeck, J.D. Gibbon and H.C. Morris, Solitons and Nonlinear Wave Equations, Academic Press, 1982.

[8] W. Eckhaus and A. van Harten, The Inverse Scattering Transformation and the Theory of Solitons, North-Holland Mathematics Studies 50, 1981 (2nd ed. 1983).

[9] C.S. Gardner, J.M. Greene, M.D. Kruskal and R.M. Miura, Method for solving the Korteweg-de Vries equation, Phys. Rev. Lett. 19 (1967), 1095-1097.

[10] C.S. Gardner, J.M. Greene, M.D. Kruskal and R.M. Miura, Korteweg-de Vries equation and generalizations VI, Comm. Pure Appl. Math. 27 (1974), 97-133.

[11] I.M. Gel'fand and B.M. Levitan, On the determination of a differential equation from its spectral function, Izvest. Akad. Nauk 15 (1951), 309-360, AMST 1 (1955), 253-309.

[12] D.J. Kaup and A.C. Newell, The Goursat and Cauchy problems for the sine-Gordon equation, SIAM J. Appl. Math. 34 (1978), 37-54.

[13] D.J. Korteweg and G. de Vries, On the change of form of long waves advancing in a rectangular canal, and on a new type of long stationary waves, Phil. Mag. 39 (1895), 422-443.

[14] G.L. Lamb, Jr., Elements of Soliton Theory, Wiley-Interscience, 1980.

[15] G.L. Lamb, Jr. and D.W. McLaughlin, Aspects of soliton physics, in: Solitons (Ed. R.K. Bullough and P.J. Caudrey) Topics in Current Physics 17, Springer, New York, 1980.

[16] P.D. Lax, Integrals of nonlinear equations of evolution and solitary waves, Comm. Pure Appl. Math. 21 (1968), 467-490.

[17] R.M. Miura, The Korteweg-de Vries equation: A survey of results, SIAM Review 18 (1976), 412-459.

[18] A.C. Scott, F.Y.F. Chu and D. McLaughlin, The soliton: a new concept in applied science, Proc. IEEE 61 (1973), 1443-1483.

[19] J. Scott Russell, Report on waves in: Report of the fourteenth meeting of the British association for the advancement of science, John Murray, London, 1844, 311-390.

[20] S. Tanaka, Modified Korteweg-de Vries equation and scattering theory, Proc. Japan Acad. 48 (1972), 466-489.

[21] S. Tanaka, On the N-tuple wave solutions of the Korteweg-de Vries equation, Publ. R.I.M.S. Kyoto Univ. 8 (1972), 419-427.

[22] S. Tanaka, Non-linear Schrödinger equation and modified Korteweg-de Vries equation; construction of solutions in terms of scattering data, Publ. R.I.M.S. Kyoto Univ. 10 (1975), 329-357.

[23] M. Wadati, The exact solution of the modified Korteweg-de Vries equation, J. Phys. Soc. Japan 32 (1972) 1681.

[24] N.J. Zabusky and M.D. Kruskal, Interactions of "solitons" in a collisionless plasma and the recurrence of initial states, Phys. Rev. Lett. 15 (1965), 240-243.

[25] V.E. Zakharov, S.V. Manakov, S.P. Novikov and L.P. Pitaievski, Theory of Solitons. The Inverse Problem Method, Nauka, Moscow, 1980 (in Russian).

[26] V.E. Zakharov and A.B. Shabat, Exact theory of two-dimensional self-focusing and one-dimensional self-modulation of waves in non-linear media, Soviet Phys. JETP (1972), 62-69.

CHAPTER ONE

THE EMERGENCE OF SOLITONS OF THE KORTEWEG-DE VRIES EQUATION FROM ARBITRARY INITIAL CONDITIONS

We study the solution $u(x,t)$ of the Korteweg-de Vries equation $u_t - 6uu_x + u_{xxx} = 0$ evolving from arbitrary real initial conditions $u(x,0) = u_0(x)$, $u_0(x)$ decaying sufficiently rapidly as $|x| \to \infty$. Using the method of the inverse scattering transformation we analyse the Gel'fand-Levitan equation in all coordinate systems moving to the right and give a complete and rigorous description of the emergence of solitons.

1. Introduction.*)

The discovery by Gardner, Greene, Kruskal and Miura [8], [9] of a method of solution for the Korteweg-de Vries equation by the inverse scattering transformation has led to a rapid and impressive development, which one can find described for example in [1], [7]. The rapid progress has produced a wealth of results, however, it has also left certain questions unanswered.

Let us recall that, by the inverse scattering transformation, the initial value problem for the (nonlinear) Korteweg-de Vries equation, is reduced to the problem of solving the (linear) Gel'fand-Levitan integral equation. The initial values $u_0(x)$, prescribed for the solution*) of the KdV equation, when introduced as a potential in the Schrödinger equation, provide the scattering data that are needed to define the kernel of the Gel'fand-Levitan equation. However, explicit solution of that equation has been obtained only in the case that the reflection coefficient corresponding to $u_0(x)$ is zero. One then has the famous "pure" N-soliton solution, with N being the number of discrete eigenvalues in the Schrödinger scattering problem.

If $u_0(x)$ is not a reflectionless potential, then by a heuristic reasoning one arrives at a conjecture about the behaviour of the solutions, as follows: one knows that solitons, if present, move to the right, while the dispersive waves that are expected to be present when the reflection coefficient is not zero, move to the left. This leads to the expectation that for large time the two ingredients of the solution will separate out, and that observers moving with suitable speeds to the right will eventually see a parade of solitons, each one followed by a decaying train of dispersive waves, as described in [10]. In spite of attempts such as [11], the question of validity of this "plausible" conjecture has not been answered in a mathematically satisfactory way.

In this chapter we study the solution $u(x,t)$ of the Korteweg-de Vries equation $u_t - 6uu_x + u_{xxx} = 0$ with arbitrary real initial conditions $u(x,0) = u_0(x)$, $u_0(x)$ decaying sufficiently rapidly as $|x| \to \infty$ for the whole of the inverse scattering transformation to hold. We analyse the Gel'fand-Levitan equation in all coordinate systems moving to the right and give a complete and rigorous description of the emergence of solitons. It will probably not be a surprise to most readers that we find N solitons emerging if $u_0(x)$ is a potential that produces N discrete eigenvalues in the Schrödinger equation, but it may be a surprise to some that this fact has never been demonstrated mathematically, except for the reflectionless potentials. We further show that the nature of decay of the dispersive wave trains behind the solitons is essentially related to properties of the reflection coefficient $b_r(k)$, at $t = 0$, such as differentiability and

*) See also the more detailed introduction to Chapter 2, especially for details about existence and uniqueness.

behaviour for $|k| \to \infty$, or the possibility of extension of $b_r(k)$ to an analytic function. These properties are in turn related to properties of the initial function $u_0(x)$.

The problem of relations between properties of potentials $u_0(x)$ and corresponding reflection coefficients $b_r(k)$ belongs to the scattering theory and is not discussed in detail in this chapter (see Chapter 2, subsection 2.1). From the literature it is known that, if $u_0(x)$ decays exponentially for $x \to +\infty$, then $b_r(k)$ can be extended to an analytic function on a strip in the upper half plane, as can be seen from [6]. Furthermore, if $u_0(x)$ and its derivatives up to fourth one decay algebraically for $|x| \to \infty$ sufficiently fast, then $b_r(k)$ belongs to $C^{(q)}(\mathbb{R})$ for some q and $b_r^{(m)}(k) = O(|k|^{-5})$ as $|k| \to \infty$, $m = 0,1,\dots,q$ (see [5]).

We also do not discuss the behaviour of solutions of the KdV equation for large time viewed in coordinate systems moving to the left, where the dispersive waves interact. Recent results on that problem (in the case of no discrete eigenvalues) have been given in [3].

The analysis given in this chapter consists of a rather simple reasoning in an abstract setting, supplemented by hard labour that is needed to obtain the necessary estimates. The reasoning is developed in sections 2 to 5, the labour is performed in section 6 and in two appendices. In the last appendix we show that our method also works in a more general setting by considering the so-called higher KdV equations.

2. Formulation of the problem.

We consider the Gel'fand-Levitan equation

$$\beta(y;x,t) + \Omega(x+y;t) + \int_0^\infty \Omega(x+y+z;t)\beta(z;x,t)dz = 0, \tag{2.1}$$

with $y > 0$, $x \in \mathbb{R}$, $t > 0$,

$$\Omega(\xi;t) = \Omega_d(\xi;t) + \Omega_c(\xi;t), \tag{2.2}$$

$$\Omega_d(\xi;t) = 2\sum_{j=1}^{N} [c_j^r(t)]^2 e^{-2\kappa_j \xi}, \quad 0 < \kappa_N < \dots < \kappa_2 < \kappa_1, \tag{2.3}$$

(2.4) $\Omega_c(\xi;t) = \frac{1}{\pi} \int_{-\infty}^{\infty} b_r(k,t)e^{2ik\xi}dk,$

(2.5) $b_r(k,t) = b_r(k)e^{8ik^3t},$

(2.6) $c_j^r(t) = c_j^r e^{4\kappa_j^3 t}.$

Here $-\kappa_j^2$, c_j^r, $j = 1,2,\ldots,N$ are the bound states and (right) normalization coefficients and $b_r(k)$ is the (right) reflection coefficient associated with the potential $u_0(x)$ in the Schrödinger scattering problem (see Chapter 2, section 2, for their definition and properties). In the integral equation (2.1) the unknown $\beta(y;x,t)$ is a function of the variable y, whereas x and t are parameters. The solution of the KdV equation is given by

(2.7) $u(x,t) = -\frac{\partial}{\partial x}\beta(0^+;x,t).$

We shall study the solution of (2.1) in moving coordinates in the parameter space x,t, defined by

(2.8) $\bar{x} = x - vt, \quad v = 4c^2, \quad c > 0.$

In particular we shall examine the behaviour for large positive times, with $\bar{x}$ confined to arbitrary half lines $\bar{x} \geq -M$, where $M \geq 0$ is independent of t. For each $c = \kappa_i$ we expect to see a soliton emerging.

We now give the problem an abstract formulation.
Let V be the Banach space of all real continuous and bounded functions g on $(0,\infty)$, equipped with the supremum norm

$$\|g\| = \sup_{0<y<+\infty} |g(y)|.$$

For $g \in V$ we write

(2.9) $(T_d g)(y) = \int_0^{\infty} \Omega_d(x+y+z;t)g(z)dz,$

(2.10) $(T_c g)(y) = \int_0^{\infty} \Omega_c(x+y+z;t)g(z)dz.$

T_d clearly is a mapping of V into V; T_c will be investigated in the next section.
Our problem is thus to find an element $\beta \in V$ such that

(2.11) $(I + T_d)\beta + T_c\beta = -\Omega,$

(2.12) $\Omega = \Omega_d + \Omega_c,$

where I is the identity mapping.
We know the solution β_d of

(2.13) $(I + T_d)\beta_d = -\Omega_d,$

which yields the pure N-soliton solution of the KdV equation. We intend to study the full problem as a perturbation of the pure N-soliton case.

3. Analysis of Ω_c and T_c.

We consider

(3.1) $$\Omega_c(\bar{x}+4c^2t+y;t) = \frac{1}{\pi}\int_{-\infty}^{\infty} b_r(k)e^{2ik(\bar{x}+y)}e^{8itk(c^2+k^2)}dk.$$

It should be clear that Ω_c is an oscillatory integral for large t, $\bar{x} \geq -M$, tending to zero as t tends to infinity. The precise behaviour depends on the behaviour of $b_r(k)$, which in turn is determined by the initial condition for the KdV equation.
Imposing suitable conditions on $b_r(k)$ we shall establish that Ω_c is strongly differentiable in V with respect to $\bar{x}$ and obeys an estimate of type

(3.2a) $|\Omega_c(\bar{x}+vt+y;t)| + |\Omega_c'(\bar{x}+vt+y;t)| \leq H(y,t),\ t \geq t_0,\ \bar{x} \geq -M,$

such that

(3.2b) $H(y,t)$ is a monotonically decreasing function of y for fixed t,

(3.2c) $$\sigma(t) \equiv \int_0^{\infty} H(z,t)dz + \sup_{0<y<+\infty} H(y,t) < +\infty,$$

(3.2d) $\sigma(t) \to 0$ as $t \to \infty$.

In (3.2a) we introduced the prime as a quick notation for the derivative with respect to $\bar{x}$.

We shall work out in detail two cases of conditions on $b_r(k)$ that are more or less typical (for a priori knowledge about $b_r(k)$ as well as the motivation for these conditions see Chapter 2, subsection 2.1).

Case C There exists an $\varepsilon > 0$ such that $b_r(k)$ is analytic on $0 < \mathrm{Im}\, k < \varepsilon$ and continuous on $0 \leq \mathrm{Im}\, k \leq \varepsilon$, while in that strip $b_r(k) = o(|k|^2)$ for $|k| \to \infty$.

Case P (i) $b_r(k)$ is $n \geq 2$ times differentiable on the real axis;

(ii) $b_r^{(m)} = o(k^{2m+2})$, $|k| \to \infty$, $m = 0,1,\dots,n-1$;

(iii) $\dfrac{b_r}{(1+|k|)^{n-1}}$, $\dfrac{b_r^{(1)}}{(1+|k|)^{n-2}}$, ..., $b_r^{(n-1)}$, $(1+|k|)b_r^{(n)}$ belong to $L^1(\mathbb{R})$.

In Case C - treated in Appendix A - one finds by means of contour integration

$$H(y,t) = \gamma e^{-2\varepsilon y} e^{-\alpha t}, \qquad \sigma(t) = O(e^{-\alpha t}), \tag{3.3}$$

where γ and α are positive constants.

In Case P - the subject of Appendix B - integration by parts produces

$$H(y,t) = \frac{\mu}{(-M+y+vt)^n}, \qquad \sigma(t) = O(\frac{1}{t^{n-1}}), \tag{3.4}$$

where again μ is a positive constant.

With the result (3.2), examplified by (3.3) and (3.4), we proceed to investigate the mapping T_c.

In moving coordinates we have

$$(T_c g)(y) = \int_0^\infty \Omega_c(\bar{x}+vt+y+z;t)g(z)dz, \tag{3.5}$$

which is a continuous function of y, since the integrand is dominated by $H(z,t)|g(z)|$. Furthermore,

$$\|T_c g\| \leq \|g\| \int_0^\infty H(z,t)dz \leq \|g\|\sigma(t). \tag{3.6}$$

We have thus established that T_c is a continuous mapping of V into V with norm tending to zero for $t \to \infty$ uniformly on $\bar{x} \geq -M$.

We next claim that T_c is strongly $\bar{x}$-differentiable in V with derivative

$$(3.7)\qquad (T_c'g)(y) = \int_0^\infty \Omega_c'(\bar{x}+vt+y+z;t)g(z)dz.$$

The proof consists in showing that

$$(3.8)\qquad A_h = \sup_{0<y<+\infty} \left| \int_0^\infty (\Delta_h(y+z)-\Omega_c'(\bar{x}+vt+y+z;t))g(z)dz \right|$$

tends to zero as $h \to 0$, where

$$(3.9)\qquad \Delta_h(y+z) = \frac{\Omega_c(\bar{x}+vt+y+z+h;t)-\Omega_c(\bar{x}+vt+y+z;t)}{h}.$$

Clearly $A_h \leq \int_0^\infty B_h(z)|g(z)|dz$, with

$$(3.10)\qquad B_h(z) = \sup_{0<y<+\infty} |\Delta_h(y+z)-\Omega_c'(\bar{x}+vt+y+z;t)|.$$

Since Ω_c is strongly $\bar{x}$-differentiable, $B_h(z)$ tends to zero as $h \to 0$. Thus in virtue of the dominated convergence theorem it suffices to show that

$$(3.11)\qquad B_h(z) \leq G(z) \text{ with } G \in L^1(0,\infty).$$

Let $\bar{x} \geq -M+\delta$, $\delta > 0$. For $|h| \leq \delta$ one has

$$(3.12)\qquad \begin{aligned} |\Delta_h(y+z)| &= |\Omega_c'(\bar{x}-\delta+vt+y+z+\theta h+\delta;t)| \text{ for some } \theta \in (0,1) \\ &\leq H(y+z+\theta h+\delta,t) \\ &\leq H(z,t). \end{aligned}$$

Consequently $B_h(z) \leq 2H(z,t)$ which is in $L^1(0,\infty)$.
Finally, we deduce from (3.2) and the above

$$(3.13)\qquad \max\left[\|\Omega_c\|,\|T_c\|,\|\Omega_c'\|,\|T_c'\|\right] \leq \sigma(t),$$

where $\sigma(t) \to 0$ as $t \to \infty$.

4. Solution of the Gel'fand-Levitan equation.

We consider

$$(4.1) \qquad (I + T_d)\beta = -(\Omega + T_c\beta).$$

Since T_d is an integral operator with degenerate kernel, solutions of

$$(I + T_d)g = f, \qquad f,g \in V$$

can be studied explicitly. In lemma 6.1 we shall show that the inverse $(I+T_d)^{-1}$ indeed exists as a mapping of V into V and that furthermore $\|(I+T_d)^{-1}\|$ is bounded for $t > 0$, $\bar{x} \geqq -M$.
To simplify the notation we shall write

$$(4.2) \qquad (I + T_d)^{-1} = S$$

and we have

$$(4.3) \qquad \|S\| \leqq a \quad \text{for } t > 0, \quad \bar{x} \geqq -M.$$

We can thus "invert" (4.1) and obtain the equation

$$(4.4) \qquad \beta = - S\Omega - ST_c\beta.$$

It can be easily shown that (4.4) possesses a unique solution $\beta \in V$. Indeed, consider the mapping $\tilde{T}$ defined by

$$(4.5) \qquad \tilde{T}g = f - ST_cg, \qquad f,g \in V.$$

By the results (3.13) and (4.3) one has

$$(4.6) \qquad \|ST_c\| \leqq \|S\|\|T_c\| \leqq a\sigma(t).$$

Hence, for sufficiently large t, we find $\|ST_c\| < 1$ and $\tilde{T}$ is a contractive mapping in the Banach space V. It follows that a unique solution g of

$$(4.7) \qquad g = f - ST_cg, \qquad f,g \in V$$

exists. Furthermore, one easily obtains an estimate for the solution. In fact, since

(4.8) $$\|g\| \leq \|f\| + \|ST_c g\| \leq \|f\| + \|ST_c\|\|g\|,$$

we obviously have

(4.9) $$\|g\| \leq \frac{1}{1 - \|ST_c\|} \|f\|.$$

5. Decomposition of the solution and estimates.

We write

(5.1) $$\beta = \beta_d + \beta_c,$$

with

(5.2) $$\beta_d = -S\Omega_d.$$

In lemma 6.2 it will be shown that $\beta_d(y;\bar{x}+vt,t)$ is uniformly bounded for $t > 0$, $\bar{x} \geq -M$, $y > 0$. We recall that β_d produces the pure N-soliton solution of the KdV equation through the formula

(5.3) $$u_d(\bar{x}+vt,t) = -\frac{\partial}{\partial \bar{x}} \beta_d(0^+;\bar{x}+vt,t).$$

Introducing the decomposition (5.1) into (4.4) we have the equation

(5.4) $$\beta_c + ST_c\beta_c = -S\Omega_c - ST_c\beta_d.$$

From the analysis of the preceding section we know that a unique solution $\beta_c \in V$ exists. To estimate the solution we proceed as follows:

(5.5) $$\|\beta_c\| \leq \|ST_c\|\|\beta_c\| + \|S\|\|\Omega_c\| + \|ST_c\|\|\beta_d\|.$$

Using (3.13), (4.3) and (4.6) one gets

(5.6) $$\|\beta_c\| \leq \frac{a\sigma(t)}{1 - a\sigma(t)} (1 + \|\beta_d\|).$$

Our final result at this stage is that in all moving coordinates $\bar{x} = x-vt$, $v > 0$, in any half line $\bar{x} \geq -M$, for large t

(5.7a) $\quad \|\beta_c\| \leq b\sigma(t),$

where b is some constant and $\sigma(t) \to 0$ as $t \to \infty$.
Furthermore, in the first approximation we have

(5.7b) $\quad \beta_c = -S(\Omega_c + T_c\beta_d) + O(\sigma^2(t)).$

Unfortunately, the labour is not finished yet. The solution of the KdV equation is given by

(5.8) $\quad u(\bar{x}+vt,t) = u_d(\bar{x}+vt,t) - \frac{\partial}{\partial \bar{x}}\beta_c(0^+;\bar{x}+vt,t).$

We thus need estimates of the derivative of β_c with respect to $\bar{x}$. To obtain these estimates we return to equation (5.4). One verifies without difficulty that both S and β_d are strongly $\bar{x}$-differentiable. From equation (5.4) we then see that β_c, too, is strongly $\bar{x}$-differentiable. Differentiating both sides with respect to $\bar{x}$ we find

(5.9) $$\beta_c' + ST_c\beta_c' = -S\{T_c'(\beta_c+\beta_d) + \Omega_c' + T_c\beta_d'\}$$
$$-S'\{\Omega_c + T_c(\beta_c+\beta_d)\}.$$

Using section 4 we again conclude that a unique solution β_c' exists and proceed to estimate the solution.
By lemma 6.2 $\beta_d'(y;\bar{x}+vt,t)$ is uniformly bounded for $t > 0$, $\bar{x} \geq -M$, $y > 0$, while lemma 6.1 tells us that

(5.10) $\quad \|S'\| \leq a \quad$ for $t > 0, \quad \bar{x} \geq -M.$

From (5.9) and the estimates (3.13), (4.3), (4.6), (5.7a) and (5.10) one finds

(5.11) $$\|\beta_c'\| \leq \frac{a\sigma(t)}{1 - a\sigma(t)}\,(2 + 2\|\beta_d\| + \|\beta_d'\| + 2b\sigma(t)).$$

Thus for large t

(5.12) $\quad \|\beta_c'\| \leq B\sigma(t),$

where B is some constant. Evidently

$$(5.13)\qquad \left|-\frac{\partial}{\partial \bar{x}}\beta_c(0^+;\bar{x}+vt,t)\right| \leq \sup_{0<y<+\infty}\left|\frac{\partial}{\partial \bar{x}}\beta_c(y;\bar{x}+vt,t)\right| = $$
$$= \|\beta_c'\|$$
$$\leq B\sigma(t).$$

We thus arrive at our final result, which can be stated as follows:

Theorem 5.1. *Let* $u(x,t)$ *be the solution of the Korteweg-de Vries problem*

$$(5.14)\qquad \begin{cases} u_t - 6uu_x + u_{xxx} = 0, & -\infty < x < +\infty, \quad t > 0 \\ u(x,0) = u_0(x), \end{cases}$$

where the real initial function $u_0(x)$ *is sufficiently smooth and decays sufficiently rapidly for* $|x| \to \infty$ *for the whole of the inverse scattering method to work and to guarantee an estimate of type (3.2). Then for any choice of the constants* $v > 0$ *and* $M \geq 0$ *there is a function* $\sigma(t)$ *such that*

$$(5.15)\qquad \sup_{x \geq -M+vt} |u(x,t) - u_d(x,t)| = \mathcal{O}(\sigma(t)) \quad \textit{as } t \to \infty.$$

Here $u_d(x,t)$ *is the pure* N*-soliton solution,* N *being the number of discrete eigenvalues corresponding to the potential* $u_0(x)$*. The function* $\sigma(t)$ *tends to zero as* $t \to \infty$ *and the exact behaviour of* $\sigma(t)$ *depends on properties of the reflection coefficient* $b_r(k)$.

If there exists an $\varepsilon > 0$ *such that* $b_r(k)$ *is analytic on* $0 < \mathrm{Im}\, k < \varepsilon$ *and continuous on* $0 \leq \mathrm{Im}\, k \leq \varepsilon$*, while in that strip* $b_r(k) = o(|k|^2)$ *for* $|k| \to \infty$*, then* $\sigma(t) = \mathcal{O}(e^{-\alpha t})$, $\alpha > 0$.

If $b_r(k)$ *is* $n \geq 2$ *times differentiable on the real axis, with* $b_r^{(m)} = o(k^{2m+2})$, $|k| \to \infty$, $m = 0,1,\ldots,n-1$ *and* $(1+|k|)^{1-n}b_r$, $(1+|k|)^{2-n}b_r^{(1)},\ldots,b_r^{(n-1)},(1+|k|)b_r^{(n)}$ *belong to* $L^1(\mathbb{R})$, *then* $\sigma(t) = \mathcal{O}(t^{1-n})$.

6. Analysis of S and β_d.

In this section we present estimates concerning S, β_d and their $\bar{x}$-derivatives that were essential in the previous investigation of the Gel'fand-Levitan equation. The original proofs as given in Chapter 2, reference [3], have been improved considerably and are therefore omitted. Instead we refer to the corresponding proofs occurring in Chapter 2.

Lemma 6.1. $I + T_d$ *is an invertible operator on the Banach space* V. *Writing*

(6.1) $$S = (I + T_d)^{-1}$$

we have

(6.2) $$\|S\|,\ \|S'\| \leq a \quad \textit{for } t > 0, \quad \bar{x} \geq -M,$$

where a is some constant and S' *denotes the strong* $\bar{x}$*-derivative of* S.

Proof: See Chapter 2, lemma 5.1.

Lemma 6.2. $\beta_d(y;\bar{x}+vt,t)$ *and* $\beta_d'(y;\bar{x}+vt,t)$ *are uniformly bounded for* $t > 0$, $\bar{x} \geq -M$, $y > 0$.

Proof: This follows from combining Chapter 2, (5.15) with Chapter 2, (5.7).

Appendix A: Case C.

Assuming that

(A.1) there exists an $\varepsilon > 0$ such that $b_r(k)$ is analytic on $0 < \text{Im } k < \varepsilon$ and continuous on $0 \leq \text{Im } k \leq \varepsilon$, while in that strip $b_r(k) = o(|k|^2)$ for $|k| \to \infty$,

we shall derive the estimate

(A.2) $$|\Omega_c(\bar{x}+4c^2t+y;t)| + |\Omega_c'(\bar{x}+4c^2t+y;t)| \leq \gamma e^{-2\varepsilon y} e^{-\alpha t}, \quad t \geq t_0,\ \bar{x} \geq -M,$$

where α and γ are positive constants.

Remark. The reason why we use no steepest descent method is that generically $b_r(k)$, if at all extendable to the upper half plane, has poles on the imaginary axis, corresponding to $i\kappa_1, i\kappa_2, \ldots, i\kappa_N$. Working as in (A.1) we avoid them.

Let us fix $c > 0$ and choose $0 < \varepsilon < c$.
Putting $w = \bar{x}+4c^2t+y$ we integrate $b_r(k)e^{2ikw}e^{8itk^3}$ around a rectangle in the complex k-plane with vertices at $-R$, R, $R+i\varepsilon$, $-R+i\varepsilon$. By (A.1) one has for $r = \pm R$

(A.3) $$\left|\int_r^{r+i\varepsilon} b_r(k)e^{2ikw}e^{8itk^3}dk\right| \leq e^{2M\varepsilon}\int_0^{\varepsilon}|b_r(r+is)|e^{-24r^2ts}ds$$
$$\leq \frac{e^{2M\varepsilon}}{24t_0}\cdot\frac{1}{r^2}\max_{0\leq s\leq\varepsilon}|b_r(r+is)|$$
$$= o(1) \text{ for } R \to \infty.$$

Thus, using Cauchy's theorem

(A.4) $$\int_{-R}^{R} b_r(k)e^{2ikw}e^{8itk^3}dk = \int_{-R}^{R} b_r(k+i\varepsilon)e^{2i(k+i\varepsilon)w}e^{8it(k+i\varepsilon)^3}dk + o(1).$$

Now the integrand on the right clearly belongs to $L^1(\mathbb{R})$.
Hence, if we let $R \to \infty$ in (A.4) we find that the integral in (3.1) is Cauchy convergent and equals

(A.5) $$\Omega_c(\bar{x}+4c^2t+y;t) =$$
$$= \frac{1}{\pi}\int_{-\infty}^{\infty} b_r(k+i\varepsilon)e^{2i(k+i\varepsilon)(\bar{x}+4c^2t+y)}e^{8it(k+i\varepsilon)^3}dk,$$

where the integral on the right converges absolutely.
From (A.5) one easily deduces

(A.6) $$|\Omega_c(\bar{x}+4c^2t+y;t)| \leq \gamma_1 e^{-2\varepsilon y}e^{-\alpha t},$$

with $\gamma_1 = \frac{1}{\pi}e^{2\varepsilon M}\int_{-\infty}^{\infty}|b_r(k+i\varepsilon)|e^{-24\varepsilon k^2t_0}dk$ and $\alpha = 8\varepsilon(c^2-\varepsilon^2) > 0$.
Obviously we can differentiate (A.5) with respect to $\bar{x}$ uniformly in y.
Estimating the derivative one finds

(A.7) $$|\Omega_c'(\overline{x}+4c^2t+y;t)| \leq \gamma_2 e^{-2\varepsilon y} e^{-\alpha t},$$

with $\gamma_2 = \frac{2}{\pi} e^{2\varepsilon M} \int_{-\infty}^{\infty} |k+i\varepsilon||b_r(k+i\varepsilon)| e^{-24\varepsilon k^2 t_0} dk.$

Hence the proof of (A.2) is complete.

Appendix B: Case P.

Let us demonstrate that the conditions

(B.1) $b_r(k)$ is $n \geq 2$ times differentiable on the real axis;

(B.2) $b_r^{(m)} = o(k^{2m+2}),\ |k| \to \infty,\ m = 0,1,\ldots,n-1;$

(B.3) $$\frac{b_r}{(1+|k|)^{n-1}},\ \frac{b_r^{(1)}}{(1+|k|)^{n-2}},\ \ldots,\ b_r^{(n-1)},\ (1+|k|)b_r^{(n)}$$

belong to $L^1(\mathbb{R})$;

guarantee the estimate

(B.4) $$|\Omega_c(\overline{x}+vt+y;t)| + |\Omega_c'(\overline{x}+vt+y;t)| \leq \frac{\mu}{(-M+y+vt)^n},$$

$$\overline{x} \geq -M, \quad t \geq t_0 > \frac{M}{v}, \quad v = 4c^2, \quad c > 0.$$

For brevity of writing we put

(B.5) $$f' = \frac{\partial}{\partial \overline{x}} f, \quad f^{(m)} = \left(\frac{\partial}{\partial k}\right)^m f, \quad f'^{(m)} = \left(\frac{\partial}{\partial k}\right)^m \frac{\partial}{\partial \overline{x}} f.$$

Furthermore we define

(B.6) $$w = \overline{x} + vt + y,$$

(B.7) $$\phi = 2k(\overline{x}+y)+8tk(c^2+k^2) = 2kw + 8tk^3,$$

(B.3) $$s = \frac{1}{\phi^{(1)}} = \frac{1}{2\overline{x}+2y+2vt+24tk^2} > 0.$$

Now, since $b_r, b_r^{(1)}, \ldots, b_r^{(n)}$ are locally integrable we can integrate by parts n times to find

(B.9) $$\int_{-R}^{R} e^{i\phi} b_r \, dk = -ise^{i\phi} \sum_{\ell=0}^{n-1} (iT)^{\ell} b_r \Big|_{-R}^{R} + \int_{-R}^{R} e^{i\phi} (iT)^n b_r \, dk,$$

where the operator T is defined by

(B.10) $$Tf = (sf)^{(1)} = s^{(1)} f + sf^{(1)}.$$

Induction reveals that the ℓ-th iterate of T has the structure

(B.11a) $$T^{\ell} f = s^{\ell} \sum_{p=0}^{\ell} \alpha_{\ell,p} f^{(\ell-p)}, \text{ with } \alpha_{\ell,0} = 1, \text{ whereas for } p \geq 1$$

(B.11b) $$\alpha_{\ell,p} = \sum_{\substack{0 \leq \ell_1, \ell_2, \ldots, \ell_p \in \mathbb{Z} \\ \ell_1 + 2\ell_2 + \ldots + p\ell_p = p}} a_{\ell;\ell_1,\ell_2,\ldots,\ell_p} \left(\frac{s^{(1)}}{s}\right)^{\ell_1} \left(\frac{s^{(2)}}{s}\right)^{\ell_2} \cdots \left(\frac{s^{(p)}}{s}\right)^{\ell_p},$$

where $a_{\ell;\ell_1,\ell_2,\ldots,\ell_p}$ are nonnegative integers, independent of s and f. Applying Leibniz' formula to the identity $(2w+24tk^2)s = 1$, we find

(B.12) $$(w+12tk^2)s^{(j)} + 24jtks^{(j-1)} + 12j(j-1)ts^{(j-2)} = 0,$$

from which it is easily seen that

(B.13) $$\left|\frac{s^{(j)}}{s}\right| \leq \frac{M_j}{(1+|k|)^j}, \quad j = 1,2,\ldots,$$

where M_j is a constant.

Thus, in view of (B.11) there are constants $A_{\ell,p}$ such that

(B.14) $$|T^{\ell} b_r| \leq s^{\ell} \sum_{p=0}^{\ell} \frac{A_{\ell,p}}{(1+|k|)^p} |b_r^{(\ell-p)}|, \quad \ell = 0,1,\ldots,n.$$

Returning to (B.9) we extract from (B.2) and (B.14) that

(B.15) $$\int_{-R}^{R} e^{i\phi} [b_r - (iT)^n b_r] dk = o(1) \text{ for } R \to \infty.$$

Hence, using (B.3), (B.14) and (3.1), we obtain

(B.16) $$\Omega_c(\overline{x}+vt+y;t) = \frac{1}{\pi} \int_{-\infty}^{\infty} e^{i\phi} (iT)^n b_r \, dk,$$

where the integral is absolutely convergent. From (B.14) we find the bound

(B.17) $$|\Omega_c(\overline{x}+vt+y;t)| \leq \frac{\mu_1}{(-M+y+vt)^n},$$

with $\mu_1 = \frac{1}{\pi 2^n} \sum_{p=0}^{n} A_{n,p} \int_{-\infty}^{\infty} \frac{|b_r^{(n-p)}|}{(1+|k|)^p} dk.$

Next, we consider the $\bar{x}$-derivative of the integrand in (B.16). Since

$$(B.18)\qquad (e^{i\phi}T^n b_r)' = 2ike^{i\phi}T^n b_r + e^{i\phi}(T^n b_r)',$$

we are led to examine $(T^n b_r)' = \sum_{p=0}^{n} (s^n \alpha_{n,p})' b_r^{(n-p)}$, in which $(s^n \alpha_{n,p})'$ is a linear combination of terms of the following form

$$(B.19)\qquad s^{\ell_0} s^{(1)\ell_1} \dots s^{(j)\ell_j - 1} s'^{(j)} \dots s^{(p)\ell_p}, \quad \ell_j \geq 1, \quad 0 \leq j \leq p,$$

where $\ell_1 + 2\ell_2 + \dots + p\ell_p = p$, $\ell_0 + \ell_1 + \dots + \ell_p = n$.

Since $s'^{(j)} = (-2s^2)^{(j)}$ we obtain from (B.13) and Leibniz' formula

$$(B.20)\qquad |s'^{(j)}| \leq \tilde{M}_j \frac{s^2}{(1+|k|)^j}, \text{ where } \tilde{M}_j \text{ is a constant.}$$

Estimating each term (B.19) by (B.13) and (B.20) one gets

$$(B.21)\qquad |(T^n b_r)'| \leq s^{n+1} \sum_{p=0}^{n} \frac{\tilde{A}_{n,p}}{(1+|k|)^p} |b_r^{(n-p)}|,$$

where the $\tilde{A}_{n,p}$ are constants.

Finally, combining (B.14), (B.18) and (B.21), we find constants $B_{n,p}$ such that

$$(B.22)\qquad |(\tfrac{1}{\pi} e^{i\phi} i^n T^n b_r)'| \leq \frac{1}{(-M+y+vt)^n} \sum_{p=0}^{n} \frac{B_{n,p}}{(1+|k|)^{p-1}} |b_r^{(n-p)}|,$$

where by (B.3) the right hand side belongs to $L^1(\mathbb{R})$.

To prove that Ω_c is strongly $\bar{x}$-differentiable we proceed as follows.

For $h > M - vt_0$ let $g(h)$ denote $T^n b_r$ with $\bar{x}$ replaced by $\bar{x} + h$. One has

$$(B.23)\qquad \left| \frac{\Omega_c(\bar{x}+h+vt+y;t) - \Omega_c(\bar{x}+vt+y;t)}{h} - \frac{i^n}{\pi} \int_{-\infty}^{\infty} (e^{i\phi} T^n b_r)' dk \right|$$

$$\leq \frac{1}{\pi |h|} \int_{-\infty}^{\infty} G_h dk, \text{ with}$$

$$(B.24)\qquad G_h = |e^{2ikh}(g(h)-g(0)-hg'(0)) + hg'(0)(e^{2ikh}-1) + $$

$$+ g(0)(e^{2ikh}-1-2ikh)|$$

$$\leq \tfrac{1}{2}h^2 |g''(\theta h)| + |hg'(0)|.|e^{2ikh}-1| + |g(0)|.|e^{2ikh}-1-2ikh|$$

for some $\theta \in (0,1)$.

Examining the $\bar{x}$-derivative of (B.19) we obtain the estimate

$$(B.25)\qquad |(T^n b_r)''| \leq s^{n+2} \sum_{p=0}^{n} \frac{\tilde{\tilde{A}}_{n,p}}{(1+|k|)^p} |b_r^{(n-p)}|,$$

where the $\tilde{\tilde{A}}_{n,p}$ are constants.
From (B.14), (B.21), (B.24) and (B.25) it is clear that

$$(B.26)\qquad \lim_{h\to 0} \frac{1}{|h|} \int_{-\infty}^{\infty} \left(\sup_{0<y<+\infty} G_h \right) dk = 0.$$

Hence, (B.23) yields that Ω_c is strongly $\bar{x}$-differentiable, the derivative being given by

$$(B.27)\qquad \Omega_c'(\bar{x}+vt+y;t) = \frac{i^n}{\pi} \int_{-\infty}^{\infty} (e^{i\phi} T^n b_r)' dk,$$

satisfying the obvious bound

$$(B.28)\qquad |\Omega_c'(\bar{x}+vt+y;t)| \leq \frac{\mu_2}{(-M+y+vt)^n},$$

$$\mu_2 = \sum_{p=0}^{n} B_{n,p} \int_{-\infty}^{\infty} \frac{|b_r^{(n-p)}|}{(1+|k|)^{p-1}} dk.$$

This completes the proof of (B.4).

Appendix C: Generalization to higher KdV equations.

We now show that the method described in this chapter is still working in a more general setting.
In [2], Appendix 3, the class of evolution equations

$$(C.1.a)\qquad q_t + C_0(L_s^+) q_x = 0, \qquad q(x,0) = q_0(x)$$

$$L_s^+ = -\tfrac{1}{4} \frac{\partial^2}{\partial x^2} - q + \tfrac{1}{2} q_x \int_x^{\infty} dy,$$

where C_0 is a ratio of entire functions, is found to be solvable by the inverse scattering transformation associated with the Schrödinger equation.

Setting $q = -u$, $C_0(z) = -\alpha(-4z)$ we can rewrite (C.1.a) as

(C.1) $$u_t = \alpha(\underline{L})u_x, \quad u(x,0) = u_0(x)$$

$$\underline{L} = \frac{\partial^2}{\partial x^2} - 4u + 2u_x \int_x^\infty dy.$$

In this form the class has been investigated by Calogero (see [4] and subsequent papers). Introducing the solution $u(x,t)$ of (C.1) as a potential in the Schrödinger equation one obtains the following simple time evolution of the spectral parameters

(C.2) $$b_r(k,t) = b_r(k)\exp\{2ik\alpha(-4k^2)t\},$$

(C.3) $$\kappa_j(t) = \kappa_j(0) \equiv \kappa_j, \quad j = 1,2,\ldots,N,$$

(C.4) $$c_j^r(t) = c_j^r\exp\{-\kappa_j\alpha(4\kappa_j^2)t\}, \quad j = 1,2,\ldots,N.$$

In particular, choosing $\alpha(z) = -z$ we rediscover the KdV equation and its wellknown time evolution in spectral space (see (2.5) and (2.6)). For simplicity we shall assume $\alpha(z)$ to be a polynomial. In this case the equations (C.1) are generally called "higher KdV equations", though, of course, this appellation is only relevant if α has degree higher than one. Let us first consider some special solutions of (C.1).

If $b_r(k) \equiv 0$ and $N = 1$, we find

(C.5) $$u(x,t) = -2\kappa_1^2\cosh^{-2}\{\kappa_1(x-\xi_0-v_1t)\},$$

(C.6) $$\xi_0 = \frac{1}{2\kappa_1}\log\left\{\frac{[c_1^r]^2}{2\kappa_1}\right\},$$

(C.7) $$v_1 = -\alpha(4\kappa_1^2),$$

which is immediately recognized as the celebrated single soliton solution.

If $b_r(k) \equiv 0$ and $N > 1$ one obtains by an exercise in linear algebra the so-called pure N-soliton solution, which is such that a transformation to moving coordinates with speed

(C.8) $$v_j = -\alpha(4\kappa_j^2), \quad j = 1,2,\ldots,N$$

makes the j-th soliton stationary as $|t| \to \infty$, provided that all v_j's are different.

In the case $b_r(k) \not\equiv 0$ dispersive waves enter in the solution. The linearized version of (C.1), reading

$$\text{(C.9)} \qquad u_t = \alpha\left(\frac{\partial^2}{\partial x^2}\right)u_x, \quad u(x,0) = u_0(x),$$

is a dispersive equation with associated group velocity

$$\text{(C.10)} \qquad v_g = -\frac{d}{dk}[k\alpha(-k^2)], \quad k \in \mathbb{R}.$$

Now, if all v_j's have the same sign, while v_g has the opposite sign (as is the case for the KdV equation with $v_j = 4\kappa_j^2$ and $v_g = -3k^2$), one expects that for large time the soliton part and the dispersive component will separate out. In order to convert this expectation into a mathematical fact we shall impose the following conditions upon the function α occurring in (C.1):

(C.11a) α is a polynomial,

(C.11b) $v_g = -\dfrac{d}{dk}[k\alpha(-k^2)] \leq 0, \quad k \in \mathbb{R},$

(C.11c) $v_j = -\alpha(4\kappa_j^2) > 0, \quad j = 1,2,\ldots,N,$

(C.11d) $v_i \neq v_j \quad$ for $i \neq j$.

To reassure the reader let us mention a class of functions meeting these requirements

$$\text{(C.12)} \qquad \alpha(z) = -z^m, \quad m \geq 1 \text{ an odd integer.}$$

For convenience we shall order the solitons so that

$$\text{(C.13)} \qquad 0 < v_N < \ldots < v_2 < v_1.$$

We can now generalize theorem 5.1 as follows:

Let u(x,t) *be the solution of the higher Korteweg-de Vries equation (C.1), with* α *as in (C.11), evolving from an arbitrary real initial function* $u_0(x)$ *(such that* $u_0(x)$ *is sufficiently smooth and decays sufficiently rapidly for* $|x| \to \infty$ *for the whole of the inverse scattering method to work and to guarantee an estimate of type (3.2)).*
Then for any choice of the constants $v > 0$ *and* $M \geq 0$ *there is a function* $\sigma(t)$ *such that*

$$\text{(C.14)} \qquad \sup_{x \geq -M+vt} |u(x,t) - u_d(x,t)| = O(\sigma(t)) \quad \text{as } t \to \infty.$$

Here $u_d(x,t)$ *is the pure* N*-soliton solution,* N *being the number of discrete eigenvalues corresponding to the potential* $u_0(x)$. *The function* $\sigma(t)$ *tends to zero as* $t \to \infty$ *and the exact behaviour of* $\sigma(t)$ *depends on properties of the reflection coefficient* $b_r(k)$.

If $b_r(k)$ *is* $n \geq 2$ *times differentiable on the real axis, with* $b_r^{(m)} = o(k^{2m+2})$, $|k| \to \infty$, $m = 0,1,\ldots,n-1$ *and* $(1+|k|)^{1-n} b_r, (1+|k|)^{2-n} b_r^{(1)}, \ldots, b_r^{(n-1)}, (1+|k|) b_r^{(n)}$ *belong to* $L^1(\mathbb{R})$, *then* $\sigma(t) = O(t^{1-n})$.

To prove this we repeat the analysis given in sections 2 to 6 and Appendix B with the following obvious adaptations:

In (2.3) the ordering is replaced by (C.13). Furthermore (2.5) and (2.6) are replaced by (C.2) and (C.4). Throughout, the speed $4c^2$ is identified with v and case C is left out.

(B.7) becomes $\phi = 2kw + 2g(k)t$; $g(k) = k\alpha(-4k^2)$.

(B.8) becomes $s = \dfrac{1}{\phi^{(1)}} = \dfrac{1}{2\bar{x}+2y+2vt+2g^{(1)}(k)t} > 0.$

$(2w+24tk^2)s = 1$ is replaced by $(2w+2g^{(1)}(k)t)s = 1$.

(B.12) becomes $(w+g^{(1)}(k)t)s^{(j)} = -t \sum_{r=1}^{j} \binom{j}{r} g^{(r+1)}(k) s^{(j-r)}$.

The reader is invited to verify in detail that our analysis remains valid once the alterations summarized above have been carried out.

References

[1] M.J. Ablowitz, Lectures on the inverse scattering transform, Stud. Appl. Math. 58 (1978), 17-94.

[2] M.J. Ablowitz, D.J. Kaup, A.C. Newell and H. Segur, The inverse scattering transform - Fourier analysis for nonlinear problems, Stud. Appl. Math. 53 (1974), 249-315.

[3] M.J. Ablowitz and H. Segur, Asymptotic solutions of the Korteweg-de Vries equation, Stud. Appl. Math. 57 (1977), 13-44.

[4] F. Calogero, A method to generate solvable nonlinear evolution equations, Lett. Nuovo Cimento 14 (1975), 443-447.

[5] A. Cohen, Existence and regularity for solutions of the Korteweg-de Vries equation, Arch. for Rat. Mech. and Anal. 71 (1979), 143-175.

[6] P. Deift and E. Trubowitz, Inverse scattering on the line, Comm. Pure Appl. Math. 32 (1979), 121-251.

[7] W. Eckhaus and A. van Harten, The Inverse Scattering Transformation and the Theory of Solitons, North-Holland Mathematics Studies 50, 1981.

[8] C.S. Gardner, J.M. Greene, M.D. Kruskal and R.M. Miura, Method for solving the Korteweg-de Vries equation, Phys. Rev. Lett. 19 (1967), 1095-1097.

[9] C.S. Gardner, J.M. Greene, M.D. Kruskal and R.M. Miura, Korteweg-de Vries equation and generalizations VI, Comm. Pure Appl. Math. 27 (1974), 97-133.

[10] R.M. Miura, The Korteweg-de Vries equation: A survey of results, SIAM Review 18 (1976), 412-459.

[11] H. Segur, The Korteweg-de Vries equation and water waves, J. Fluid Mech. 59 (1973), 721-736.

CHAPTER TWO

ASYMPTOTIC ESTIMATES OF SOLUTIONS OF THE KORTEWEG-DE VRIES EQUATION ON RIGHT HALF LINES SLOWLY MOVING TO THE LEFT

We consider the Korteweg-de Vries equation $u_t - 6uu_x + u_{xxx} = 0$ with arbitrary real initial conditions $u(x,0) = u_0(x)$, sufficiently smooth and rapidly decaying as $|x| \to \infty$. Using the method of the inverse scattering transformation we analyse the behaviour of the solution $u(x,t)$ in coordinate regions of the form $t \geq t_0$, $x \geq -\mu - \nu T$, $T = (3t)^{1/3}$ where μ, ν and t_0 are nonnegative constants. We derive explicit x and t dependent bounds for the nonsoliton part of $u(x,t)$. These bounds enable us to prove a convergence result, which clearly displays the emergence of solitons. Furthermore, they help us to establish some interesting momentum and energy decomposition formulae.

1. Introduction.

We study the Korteweg-de Vries (KdV) problem

(1.1a) $$u_t - 6uu_x + u_{xxx} = 0, \quad -\infty < x < +\infty, \quad t > 0$$

(1.1b) $$u(x,0) = u_0(x),$$

where the initial function $u_0(x)$ is an arbitrary real function on $\mathbb{R}$, such that

(1.2) $u_0(x)$ is sufficiently smooth and (along with a number of its derivatives) decays sufficiently rapidly for $|x| \to \infty$ for the whole of the inverse scattering method to work and to guarantee certain regularity and decay properties of the right reflection coefficient, to be stated further on.

To make the discussion less abstract let us quote a definite example from [4] in which (1.2) is fulfilled:

(1.3a) u_0 is of class C^3 on $\mathbb{R}$ and has a piecewise continuous fourth derivative,

(1.3b) $u_0^{(j)}(x) = O(|x|^{-M})$ as $|x| \to \infty$ for $j \leq 4$,

where $M > \gamma$. Here γ is a constant which is 8 in the generic case (see (2.13)) but which is 10 in the exceptional case (see (2.14)). In [4] it is shown by an inverse scattering analysis that condition (1.3) guarantees the existence of a real function $u(x,t)$, continuous on $\mathbb{R}\times[0,\infty)$, which satisfies (1.1) in the classical sense.

Let us recall that there is uniqueness of solutions of (1.1) within the "Lax-class", i.e. the class of functions which, together with a sufficient number of derivatives vanish for $|x| \to \infty$, as discussed in [12]. Whenever, in the sequel, we speak of "the solution" of (1.1) we shall refer to the solution obtained by inverse scattering (unique within the Lax-class).

The long-time behaviour of the solution $u(x,t)$ of (1.1) has been discussed by several authors, the general picture being, that as $t \to \infty$ the solution decomposes into a finite number of solitons moving to the right and a dispersive wavetrain moving to the left. The emergence of solitons from initial conditions as arbitrary as (1.2) was demonstrated rigorously in [8] corresponding to Chapter 1 of this volume. It was proven there, that for any choice of the constants $v > 0$ and $M \geq 0$ there is a function $\sigma(t)$ such that

(1.4) $$\sup_{x \geq -M+vt} |u(x,t) - u_d(x,t)| = O(\sigma(t)) \text{ as } t \to \infty,$$

where $u_d(x,t)$ denotes the pure N-soliton solution (see (5.16)), N being the number of bound states corresponding to $u_0(x)$, when introduced as a potential in the Schrödinger scattering problem. The function $\sigma(t)$ tends to zero as $t \to \infty$ and the exact behaviour of $\sigma(t)$ depends on properties of u_0.

If u_0 decays exponentially for $x \to +\infty$ then $\sigma(t) = \mathcal{O}(e^{-\alpha t})$ for some constant $\alpha > 0$.

If u_0 and its derivatives up to the fourth one decay algebraically for $|x| \to \infty$ sufficiently fast, then $\sigma(t) = \mathcal{O}(t^{-m})$ for some constant $m \geq 1$. Earlier results in this direction, though less detailed and not widely known, were given in [16], where it was shown that

(1.5) $$\lim_{t\to\infty} \sup_{x>vt} |u(x,t) - u_d(x,t)| = 0 \text{ for } v > 0 \text{ arbitrarily fixed.}$$

In fact, in recent years, most of the asymptotic attention was devoted to the solitonless KdV initial value problem. In this case, the analysis given in [1] led to the recognition of four distinct asymptotic regions

I. $x \geq \tilde{\mathcal{O}}(t)$ II. $|x| \leq \tilde{\mathcal{O}}(t^{1/3})$
III. $-x = \tilde{\mathcal{O}}\{t^{1/3}(\ln t)^{2/3}\}$ IV. $-x \geq \tilde{\mathcal{O}}(t)$,

where $\tilde{\mathcal{O}}$ denotes positive proportionality. Within each region, the solution u(x,t) has an asymptotic expansion, characteristic for that region. However, interesting as they may be, the results are far from rigorous. Indeed, discussing the matter in their book [2], p. 68 the authors remark

> "Two warnings should be made before we begin the analysis. The first is that almost none of the results to be described in this section are known rigorously. These results are formal, and have great practical value, but proofs of asymptoticity are yet to be given. The second (related) warning is that some of the existing literature on this question contains errors".

In this chapter we extend the asymptotic analysis given in Chapter 1. We use the method of the inverse scattering transformation to analyse the behaviour of the solution u(x,t) of (1.1) - both in the absence of solitons as well as in the general case - in coordinate regions of the form

(1.6) $t \geq t_0, \quad x \geq -\zeta, \quad \zeta = \mu + \nu T, \quad T = (3t)^{1/3},$

where μ, ν and t_0 are nonnegative constants. Here μ is arbitrary, but the values of ν are restricted to $0 \leq \nu < \nu_c$ where ν_c is some generic number connected with the Airy function, its numerical value being 1.39. Furthermore, t_0 depends on μ, ν and the behaviour of u_0 as well. Note that (1.6) covers region I and almost all of region II.
It is shown that in the absence of solitons

(1.7) $\sup_{x \geq -\zeta} |u(x,t)| = O(t^{-2/3})$ as $t \to \infty$.

In the general case we improve (1.5) by

(1.8) $\sup_{x \geq -\zeta} |u(x,t) - u_d(x,t)| = O(t^{-1/3})$ as $t \to \infty$.

This leads to the convergence result (7.17), which clearly displays the emergence of solitons. Moreover, we construct several remarkably explicit x and t dependent bounds for the nonsoliton part of the solution valid in regions (1.6). With the help of these bounds we establish the momentum and energy decomposition formulae

(1.9a) $\int_{-\zeta}^{\infty} u(x,t)dx = -4 \sum_{p=1}^{N} \kappa_p + O(t^{-1/3})$ as $t \to \infty$

(1.9b) $\int_{-\infty}^{-\zeta} u(x,t)dx = -\frac{2}{\pi} \int_0^{\infty} \log(1-|b_r(k)|^2)dk + O(t^{-1/3})$ as $t \to \infty$

(1.9c) $\int_{-\zeta}^{\infty} u^2(x,t)dx = \frac{16}{3} \sum_{p=1}^{N} \kappa_p^3 + O(t^{-1/3})$ as $t \to \infty$

(1.9d) $\int_{-\infty}^{-\zeta} u^2(x,t)dx = -\frac{8}{\pi} \int_0^{\infty} k^2 \log(1-|b_r(k)|^2)dk + O(t^{-1/3})$ as $t \to \infty$,

where $-\kappa_p^2$ ($\kappa_p > 0$) $p = 1,2,\ldots,N$ are the bound states and $b_r(k)$ is the right reflection coefficient associated with the potential $u_0(x)$ in the Schrödinger scattering problem.

Let us point out three major differences between our approach and [1]. Firstly, it is our main purpose to obtain explicit bounds for the nonsoliton part of the solution of (1.1), valid in the region (1.6), whereas in [1] the emphasis lies on the construction of asymptotic expansions of the (solitonless) solution in the various regions. Secondly, the analysis in [1] requires that u_0 decays faster than exponentially as $x \to +\infty$. For our results, however, except explicitly stated otherwise, an algebraic decay rate of type (1.3) is sufficient. The third difference lies in the fact that we venture to call our results rigorous.

The composition of this chapter is as follows. In section 2 we briefly discuss the direct and inverse scattering problem for the Schrödinger equation and formulate our problem. In section 3 we isolate certain properties of the function Ω_c and the operator T_c, both occurring in the Gel'fand-Levitan equation. These properties are used in section 4 to investigate the solution of (1.1) in the absence of solitons. In the subsequent sections it is supposed that solitons are present. In that case the operator $I + T_d$ appears in the Gel'fand-Levitan equation. Section 5 is devoted to showing that this operator is invertible. In section 6, the operator $S = (I + T_d)^{-1}$ is applied to both sides of the Gel'fand-Levitan equation. It is shown that the resulting equation has a unique solution β, representable by a Neumann series. Finally, in section 7, we write β as the N-soliton state plus a perturbation. This leads to a decomposition of the solution $u(x,t)$ of (1.1) into an N-soliton part and a nonsoliton part. For the nonsoliton part we derive explicit estimates and discuss their consequences.

The notation used is similar to that in Chapter 1. Since the constants appearing in the sequel have a simple structure, we have traced them all. In this way the interested reader can obtain numerical estimates in a practical case.

2. Preliminaries and statement of the problem.

2.1. Direct scattering at $t = 0$.

Let us briefly review the direct scattering problem for the Schrödinger equation

$$(2.1) \qquad \psi_{xx} + (k^2 - u_0(x))\psi = 0, \qquad -\infty < x < +\infty,$$

where u_0 is the initial function in (1.1b) and k a complex parameter. For details we refer to [4], [6] and [7]. Our notation closely resembles that used in [7].

The results of this subsection are valid for any real function $u_0(x)$, continuous on $\mathbb{R}$, vanishing at infinity and integrable with respect to $x^2\,dx$.

For Im k ≥ 0 we introduce the Jost functions $\psi_r(x,k)$ and $\psi_\ell(x,k)$, two special solutions of (2.1) uniquely determined by

$$\text{(2.2a)} \qquad \psi_r(x,k) = e^{-ikx}R(x,k), \quad \lim_{x\to-\infty} R(x,k) = 1, \quad \lim_{x\to-\infty} R_x(x,k) = 0$$

$$\text{(2.2b)} \qquad \psi_\ell(x,k) = e^{ikx}L(x,k), \quad \lim_{x\to+\infty} L(x,k) = 1, \quad \lim_{x\to+\infty} L_x(x,k) = 0.$$

The functions R, R_x, R_{xx}, L, L_x, L_{xx} are continuous in (x,k) on $\mathbb{R}\times\overline{\mathbb{C}}_+$ and analytic in k on $\mathbb{C}_+$ for each $x \in \mathbb{R}$. Furthermore, for k real, $k \neq 0$, the pairs $\psi_\ell(x,k)$, $\psi_\ell(x,-k)$ and $\psi_r(x,k)$, $\psi_r(x,-k)$ constitute fundamental systems of solutions of equation (2.1).
In particular we have for $x \in \mathbb{R}$, $k \in \mathbb{R}\backslash\{0\}$

$$\text{(2.3a)} \qquad \psi_r(x,k) = r_+(k)\psi_\ell(x,k) + r_-(k)\psi_\ell(x,-k)$$

$$\text{(2.3b)} \qquad r_+(k) = (2ik)^{-1}W[\psi_\ell(x,-k),\psi_r(x,k)]$$

$$\text{(2.3c)} \qquad r_-(k) = (2ik)^{-1}W[\psi_r(x,k),\psi_\ell(x,k)],$$

where $W[\psi_1,\psi_2] = \psi_1\psi_{2x} - \psi_{1x}\psi_2$ denotes the Wronskian of ψ_1 and ψ_2. It is easily verified that for $k \in \mathbb{R}\backslash\{0\}$, writing * for complex conjugation:

$$\text{(2.4)} \qquad r_+^*(k) = r_+(-k) \qquad\qquad r_-^*(k) = r_-(-k)$$

$$\text{(2.5)} \qquad |r_-(k)|^2 = |r_+(k)|^2 + 1.$$

The representation (2.3c) enables us to extend $r_-(k)$ to a function analytic on Im k > 0 and continuous on Im k ≥ 0, k ≠ 0. One can prove, that $r_-(k)$ has at most finitely many zeros, all simple and on the imaginary axis. Let us denote them by $i\kappa_j$, j = 1,2,...,N and order

$$\text{(2.6)} \qquad \kappa_1 > \kappa_2 > \dots > \kappa_N > 0.$$

It turns out that the eigenvalues of (2.1) (the so-called bound states) are given by $-\kappa_1^2 < -\kappa_2^2 < \dots < -\kappa_N^2$. The associated L^2-eigenspaces are one-dimensional and spanned by the real-valued exponentially decaying functions $\psi_\ell(x,i\kappa_j)$, $j = 1,2,\dots,N$. In terms of these one defines the (right) normalization coefficients

$$(2.7) \qquad c_j^r = \left[\int_{-\infty}^{\infty} \psi_\ell^2(x,i\kappa_j)dx\right]^{-\frac{1}{2}}.$$

Note that

$$(2.8) \qquad c_j^r = \lim_{x\to+\infty} \psi_j(x)e^{\kappa_j x},$$

where $\psi_j(x)$ stands for the normalized eigenfunction $\|\psi_\ell(.,i\kappa_j)\|_{L^2}^{-1}\psi_\ell(x,i\kappa_j)$. In [6] it is proven that (2.1) has a bound state if and only if $\psi_\ell(x,0)$ vanishes for some x.

Next, we introduce the following functions for $k \in \mathbb{R}\backslash\{0\}$

$$(2.9a) \qquad a_r = r_-^{-1}, \text{ the (right) transmission coefficient}$$

$$(2.9b) \qquad b_r = r_+r_-^{-1}, \text{ the (right) reflection coefficient.}$$

The appellation is motivated by the asymptotic behaviour of ψ_r for $|x| \to \infty$ with $k \in \mathbb{R}\backslash\{0\}$ fixed:

$$(2.10) \qquad a_r(k)\psi_r(x,k) \approx e^{-ikx} + b_r(k)e^{ikx} \qquad \text{as } x \to +\infty$$

$$\approx a_r(k)e^{-ikx} \qquad \text{as } x \to -\infty.$$

The growth of u_0 permits us to extend a_r and b_r in a natural way to continuous functions on all of $\mathbb{R}$, where they satisfy

$$(2.11) \qquad a_r^*(k) = a_r(-k), \qquad b_r^*(k) = b_r(-k)$$

$$(2.12) \qquad |a_r(k)|^2 + |b_r(k)|^2 = 1 \text{ and for } k \in \mathbb{R}\backslash\{0\}\ |a_r(k)| > 0.$$

It is shown in [7], that b_r is an element of $L^1 \cap L^2(\mathbb{R})$, which behaves as $o(|k|^{-1})$ for $k \to \pm\infty$.

We shall call the aggregate of quantities $\{b_r(k),\kappa_j,c_j^r\}$ the (right) scattering data associated with the potential u_0. It is a remarkable fact that a potential is completely determined by its scattering data.

In general, starting from a given potential u_0, it is not possible to obtain the scattering data in closed form. An exception constitutes the potential $u_0(x) = -\lambda(\lambda + 1)\mathrm{sech}^2 x$, $\lambda > 0$, as can be seen from Chapter 3, section 5.

Let us examine the behaviour of the reflection coefficient $b_r(k)$ in some detail. Following [4], we shall from now on distinguish two cases, the "generic case" and the "exceptional case". In the generic case, the Jost functions $\psi_r(x,0)$ and $\psi_\ell(x,0)$ are linearly independent

$$(2.13)\qquad \lim_{\substack{k\to 0\\ \mathrm{Im}\,k\geq 0}} 2ikr_-(k) = W[\psi_r(x,0),\psi_\ell(x,0)] \neq 0,$$

whereas the exceptional case is characterized by

$$(2.14)\qquad \lim_{\substack{k\to 0\\ \mathrm{Im}\,k\geq 0}} 2ikr_-(k) = W[\psi_r(x,0),\psi_\ell(x,0)] = 0,$$

expressing the linear dependence of $\psi_r(x,0)$ and $\psi_\ell(x,0)$. The value of b_r at $k = 0$ will frequently appear in our analysis. Therefore, let us give it due attention.

In the generic case, we obtain from (2.3-13)

$$(2.15)\qquad b_r(k) = -1 + \alpha_r k + o(|k|) \text{ as } k \to 0,$$

where $\alpha_r \neq 0$ is some constant.

In the exceptional case, there is a constant $\alpha_0 \in \mathbb{R}\backslash\{0\}$ such that

$$(2.16)\qquad \psi_r(x,0) = \alpha_0\psi_\ell(x,0).$$

As a consequence of (2.3-14) both r_+ and r_- have a finite limit as $k \to 0$. Taking $k \to 0$ in (2.3a) we find

$$(2.17)\qquad r_+(0) + r_-(0) = \alpha_0.$$

Hence, in view of (2.4-5)

$$(2.18)\qquad r_+(0) = \tfrac{1}{2}(\alpha_0 - \alpha_0^{-1}), \qquad r_-(0) = \tfrac{1}{2}(\alpha_0 + \alpha_0^{-1}),$$

$$(2.19)\qquad b_r(0) = \frac{\alpha_0^2 - 1}{\alpha_0^2 + 1} \in (-1,1).$$

Observe that $b_r(0) = 0$ if and only if we are in the "degenerate exceptional case" characterized by

(2.20) $\psi_r(x,0) = \pm\psi_\ell(x,0)$.

As an example of an exceptional case consider the potential $u_0 = v_x + v^2$, $v \in C^1(\mathbb{R})$, where v and v_x decay rapidly enough. Then

(2.21a) $\psi_r(x,0) = \exp\left[\int_{-\infty}^{x} v(y)dy\right]$, $\psi_\ell(x,0) = \exp\left[-\int_{x}^{\infty} v(y)dy\right]$

(2.21b) $\alpha_0 = \exp\left[\int_{-\infty}^{\infty} v(y)dy\right]$, $\quad b_r(0) = \tanh\left[\int_{-\infty}^{\infty} v(y)dy\right]$.

Since $\psi_\ell(x,0)$ does not vanish, there are no bound states in this example. On the other hand, any potential $u_0 \geqq 0$, $u_0 \not\equiv 0$, is generic with no bound states.

In [13] it was supposed that the zeros of $b_r(k)$ are associated with the discontinuities of $u_0(x)$ and that $b_r(k)$ would have no zeros for a perfectly smooth initial condition. The previous example shows that this supposition is incorrect: e.g., any rapidly decaying odd function v of class $C^\infty(\mathbb{R})$ produces a perfectly smooth u_0 with $b_r(0) = 0$.

Next, let us mention some regularity and decay properties of $b_r(k)$. If $u_0(x)$ merely satisfies the conditions stated at the beginning of this subsection, then we have $b_r(k) = o(|k|^{-1})$ for $k \to \pm\infty$ and no better. However, if $u_0(x)$ has rapidly decaying derivatives, then $b_r(k)$ has rapidly decaying derivatives as well. The converse is also true. Specifically, u_0 is in the Schwartz class if and only if b_r is in the Schwartz class.

To be less demanding, suppose u_0 satisfies the hypotheses (1.3). Then [4], lemma 2.4 tells us

(2.22a) In the generic case, b_r is of class $C^{[\![M]\!]-2}$ on $\mathbb{R}$.

(2.22b) In the exceptional case, b_r is of class $C^{[\![M]\!]-2}$ on $\mathbb{R}\setminus\{0\}$ and of class $C^{[\![M]\!]-3}$ on all of $\mathbb{R}$.

(2.22c) In either case, $b_r^{(j)}(k) = O(|k|^{-5})$ as $k \to \pm\infty$ for $j < M-2$.

Here $[\![M]\!]$ denotes the largest integer strictly less than M. As a final remark, let us mention that it is straightforward to obtain the following result from the representation formulae (4.2.4.2-4) in [7]:

If u_0 satisfies (1.3) and there exists an $\varepsilon_0 > 0$ such that $u_0(x) = O(\exp(-2\varepsilon_0 x))$ as $x \to +\infty$ then for any ε_1 with $0 < \varepsilon_1 < \min(\varepsilon_0, \kappa_N)$ the function $b_r(k)$ is analytic on $0 < \operatorname{Im} k < \varepsilon_1$ and continuous on $0 \leq \operatorname{Im} k \leq \varepsilon_1$, while in that strip $b_r(k) = O(|k|^{-1})$ as $|k| \to \infty$. In particular, if u_0 decays faster than exponentially as $x \to +\infty$ and has no bound states, then $b_r(k)$ is analytic on the open upper half plane.

In the derivation of the asymptotic expansions given in [1] the extension of b_r to the upper half plane plays a crucial role. It is therefore surprising that we shall not need any supposition of this kind for our asymptotic analysis, except for an example given at the end of section 7.

2.2. Inverse scattering for $t > 0$.

From now on we shall impose stronger conditions on the initial function u_0, than those stated at the beginning of the preceding subsection. In fact, we shall assume that (1.2) is fulfilled.

As a specific example it is good to keep in mind that, except explicitly stated otherwise, all the results of this paper are valid if (1.3) holds.

Consequently, there is a unique (within the Lax-class) real function $u(x,t)$, continuous on $\mathbb{R}\times[0,\infty)$ and satisfying (1.1b), which is a classical solution of the KdV equation (1.1a) for all $t > 0$.

Moreover, for fixed $t \geq 0$, the function $u(x,t)$ decays rapidly enough for $|x| \to \infty$ to fall in the class of potentials discussed in the previous subsection.

The important discovery by Gardner-Greene-Kruskal-Miura [9], [10] is the following:

For each $t \geq 0$, introduce $u(x,t)$ as a potential in the Schrödinger scattering problem

$$(2.23) \qquad \psi_{xx} + (k^2 - u(x,t))\psi = 0, \qquad -\infty < x < +\infty;$$

then the bound states $-\kappa_1^2 < -\kappa_2^2 < \ldots < -\kappa_N^2$ do not change with time, whereas the associated normalization coefficients and reflection coefficient change in a simple way

$$(2.24) \qquad c_j^r(t) = c_j^r \exp\{4\kappa_j^3 t\}, \qquad j = 1,2,\ldots,N$$

(2.25) $$b_r(k,t) = b_r(k)\exp\{8ik^3t\}, \qquad -\infty < k < +\infty.$$

To determine the solution of the KdV problem for all $t > 0$, one exploits the fact that the potential $u(x,t)$ can be recovered from the scattering data $\{b_r(k,t),\kappa_j,c_j^r(t)\}$ by solving the inverse scattering problem. For that purpose we introduce the real functions

(2.26a) $$\Omega(\xi;t) = \Omega_d(\xi;t) + \Omega_c(\xi;t), \qquad \xi \in \mathbb{R}, \quad t > 0$$

(2.26b) $$\Omega_d(\xi;t) = 2\sum_{j=1}^{N} [c_j^r(t)]^2 e^{-2\kappa_j\xi}, \qquad 0 < \kappa_N < \ldots < \kappa_2 < \kappa_1$$

(2.26c) $$\Omega_c(\xi;t) = \frac{1}{\pi}\int_{-\infty}^{\infty} b_r(k,t)e^{2ik\xi}dk.$$

Since $b_r(k,t)$ is in $L^1 \cap C_0$ $(-\infty < k < +\infty)$, the integral in (2.26c) converges absolutely and $\Omega_c(\xi;t)$ belongs to $L^2 \cap C_0$ $(-\infty < \xi < +\infty)$. Consider now the Gel'fand-Levitan equation (see [7])

(2.27) $$\beta(y;x,t) + \Omega(x+y;t) + \int_0^{\infty} \Omega(x+y+z;t)\beta(z;x,t)dz = 0,$$

with $y > 0$, $x \in \mathbb{R}$, $t > 0$.
In this integral equation the unknown $\beta(y;x,t)$ is a function of the variable y, whereas x and t are parameters. For each $x \in \mathbb{R}$, $t > 0$ there is a unique solution $\beta(y;x,t)$ to (2.27) in L^2 $(0 < y < +\infty)$. Furthermore, $\beta(y;x,t)$ is real and belongs to $L^1 \cap L^2 \cap C$ $(0 < y < +\infty)$, such that both limits at the boundary of $0 < y < +\infty$ exist. In fact we have

(2.28a) $$\lim_{y\to+\infty} \beta(y;x,t) = 0$$

as well as the important property

(2.28b) $$\beta(0^+;x,t) = \int_x^{\infty} u(\tilde{x},t)d\tilde{x}, \qquad x \in \mathbb{R}, \quad t > 0.$$

Herewith the inverse scattering problem is solved, since the solution of the KdV problem is given by

(2.29) $$u(x,t) = -\frac{\partial}{\partial x}\beta(0^+;x,t), \qquad x \in \mathbb{R}, \quad t > 0.$$

Note that the original problem for the nonlinear partial differential equation (1.1) is essentially reduced in this way to the problem of solving a one-dimensional linear integral equation. Explicit solutions of (2.27)

have only been obtained for $b_r \equiv 0$. On account of its asymptotic behaviour (cf. (5.21)), the solution $u_d(x,t)$ of the KdV equation with scattering data $\{0,\kappa_j,c_j^r(t)\}$ is called the pure N-soliton solution associated with $u_0(x)$.

2.3. Statement of the problem.

We shall study the solution of (2.27) in parameter regions of the form

$$(2.30) \qquad t \geq t_0, \quad x \geq -\mu - \nu T, \quad T = (3t)^{1/3}$$

where μ, ν and t_0 are nonnegative constants. Here μ is arbitrary, but the values of ν are restricted to $0 \leq \nu < \nu_c$ where ν_c is some generic number to be specified later on. Furthermore t_0 depends on μ, ν and u_0.

It is essential to give our problem a convenient abstract formulation. Remarkably enough the function space introduced in Chapter 1 also works in the more general setting of this chapter.

So, once again, let V denote the Banach space of all real continuous and bounded functions g on $(0,\infty)$, equipped with the supremum norm

$$\|g\| = \sup_{0<y<+\infty} |g(y)|$$

and write

$$(2.31) \qquad (T_d g)(y) = \int_0^\infty \Omega_d(x+y+z;t)g(z)dz$$

$$(2.32) \qquad (T_c g)(y) = \int_0^\infty \Omega_c(x+y+z;t)g(z)dz.$$

As in Chapter 1, it is readily verified that T_d maps V into V. In the next section it is shown that if the right reflection coefficient $b_r(k)$ satisfies certain, rather modest regularity and decay conditions, then T_c is indeed a mapping of V into V. Nevertheless, there is a fundamental difference with the situation of Chapter 1. There, $\|T_c\|$ tends to zero as $t \to \infty$ (cf. Chapter 1,(3.13)). In the region (2.30), however, T_c is only a small operator. In fact, ν_c must be chosen with due care so as to guarantee that $\|T_c\| < 1$. More so, it is an amusing question whether T_c is a small operator in the region (2.30) with ν arbitrary! Plainly, this difference in the long-time behaviour of $\|T_c\|$ has its impact on the technicalities (see section 6).

In the above notation our problem amounts to analysing the solution of

(2.33a) $\quad (I + T_d)\beta + T_c\beta = -\Omega, \qquad \beta \in V$

(2.33b) $\quad \Omega = \Omega_d + \Omega_c,$

- where I is the identity mapping - in the parameter region (2.30).
As in Chapter 1, we know the solution β_d of

(2.34) $\quad (I + T_d)\beta_d = -\Omega_d,$

yielding the pure N-soliton solution of the KdV equation. Despite the differences in the long-time behaviour of $\|T_c\|$, the basic thought of the analysis in this chapter is the same as in Chapter 1, namely to treat the full problem (2.33) as a perturbation of the pure N-soliton case (2.34).

3. Analysis of Ω_c and T_c.

Let us examine

(3.1) $$\Omega_c(x+y;t) = \frac{1}{\pi} \int_{-\infty}^{\infty} b_r(k)e^{2ik(x+y)+8ik^3t}dk$$

in the parameter region $T = (3t)^{1/3} > 0$, $x \geq -\mu - \nu T$, where μ and ν are nonnegative constants. Performing the change of variables $s = 2kT$, $T = (3t)^{1/3}$ we find

(3.2) $$\Omega_c(x+y;t) = \frac{1}{2\pi T} \int_{-\infty}^{\infty} b_r(\frac{s}{2T})\exp\left[i(\frac{x+y}{T})s + i\,\frac{s^3}{3}\right]ds,$$

which shows that the asymptotic behaviour of Ω_c for $t \to \infty$ is intimately related to the behaviour of the Airy function

(3.3) $$Ai(\eta) = \frac{1}{2\pi} \lim_{R\to\infty} \int_{-R}^{R} \exp[i\eta s + i\,\frac{s^3}{3}]ds, \qquad \eta \in \mathbb{R}.$$

Of course we can rewrite (3.3) as

(3.4) $$Ai(\eta) = \frac{1}{\pi} \lim_{R\to\infty} \operatorname{Re} \int_0^R \exp[i\eta s + i\,\frac{s^3}{3}]ds$$

$$= \frac{1}{\pi} \lim_{R\to\infty} \int_0^R \cos[\eta s + \frac{s^3}{3}]ds, \qquad \eta \in \mathbb{R}$$

a form which is also frequently encountered in the literature.
Let us list some wellknown [14] properties of (3.3) that will be used in the sequel. The function $Ai(\eta)$ has an analytic continuation to the whole complex η plane, which satisfies

(3.5) $$\frac{d^2}{d\eta^2} Ai(\eta) = \eta\, Ai(\eta), \qquad \eta \in \mathbb{C}.$$

For $\eta > 0$ one has the estimates

(3.6a) $$0 < Ai(\eta) \le \frac{e^{-\chi}}{2\pi^{1/2}\eta^{1/4}}, \quad 0 < -\frac{d}{d\eta}Ai(\eta) \le \frac{\eta^{1/4}e^{-\chi}}{2\pi^{1/2}}(1 + \frac{7}{72\chi}), \text{ with}$$

(3.6b) $$\chi = \frac{2}{3}\eta^{3/2}.$$

Together, (3.5) and (3.6) imply that for any constant $\nu \geqq 0$

(3.7) $$a_{m,\theta}(\nu) \equiv \sup_{\eta\geqq-\nu} \left\{|Ai^{(m)}(\eta)|\exp\left[\frac{2}{3}\theta(\nu + \eta)^{3/2}\right]\right\} < +\infty$$

for $0 \leqq \theta < 1$, $m = 0,1,2,\ldots$.
Let us conclude with an innocent looking property of the Airy function that plays a key role in this chapter

(3.8) $$\int_0^\infty Ai(\eta)d\eta = \frac{1}{3}.$$

Starting point for our analysis of Ω_c in the parameter region $T = (3t)^{1/3} > 0$, $x \geqq -\mu - \nu T$ is the following transcription of (3.2)

(3.9a) $$\Omega_c(x+y;t) = \frac{\varepsilon}{\pi} \int_{-\infty}^{\infty} b(\varepsilon s,\mu)\exp[i\eta s + i\,\frac{s^3}{3}]ds, \text{ with}$$

(3.9b) $$b(k,\mu) = b_r(k)e^{-2ik\mu}, \qquad \varepsilon = \frac{1}{2T}, \qquad \eta = \frac{x+y+\mu}{T}.$$

Thanks to this representation our task is reduced to the examination of the integral (3.9a) in the parameter region $\varepsilon > 0$, $\eta \geqq -\nu$.
Basically, this examination is performed in the next two lemmas. These are then combined to give theorem 3.3 describing the structure of Ω_c.
Throughout the following notational convention is used. If $f : \mathbb{R}\times[0,\infty) \to \mathbb{C}$ is bounded, then we write for $\mu \geqq 0$

$$\|f\|_\infty = \|f\|_\infty(\mu) = \sup_{k\in\mathbb{R}} |f(k,\mu)|.$$

The first lemma isolates certain regularity and decay properties with respect to η of integrals of type (3.9a).

Lemma 3.1. *Let* g *be of class* $C^2[0,\infty)$. *Assume that the derivatives* $g^{(j)}(s)$, $j = 0,1,2$ *satisfy*

$$(3.10) \qquad g(s) = O(s), \quad g^{(1)}(s) = O(s), \quad g^{(2)}(s) = O(1)$$

for $s \to +\infty$. *Then*

$$(3.11) \qquad I(\eta) = \lim_{R\to\infty} \int_0^R g(s)\exp[i\eta s + i\frac{s^3}{3}]ds$$

is well-defined for all $\eta \in \mathbb{R}$ *and* $I \in C^2$ $(-\infty < \eta < +\infty)$ *with*

$$(3.12) \qquad \lim_{\eta\to+\infty} (\frac{d}{d\eta})^j I(\eta) = 0, \quad j = 0,1,2.$$

Furthermore, if instead of (3.10) we make the stronger assumption

$$(3.13) \qquad g^{(j)}(s) = O(1) \quad \text{for } s \to +\infty$$

then the following representation holds

$$(3.14) \qquad \frac{d}{d\eta} I(\eta) = \lim_{R\to\infty} \int_0^R isg(s)\exp[i\eta s + i\frac{s^3}{3}]ds.$$

Proof: The idea of the proof is to rewrite $I(\eta)$ as a nice integral over a finite interval plus a remainder which can be treated using integration by parts.

In fact, we claim the following. Let $\eta_0 \geq 0$ be an arbitrary constant. Set $s_0 = \sqrt{1 + \eta_0}$. Then for $\eta \geq -\eta_0$ one has

$$(3.15) \qquad I(\eta) = I_1(\eta) + I_2(\eta) + I_3(\eta), \quad \text{with}$$

$$(3.16a) \qquad I_1(\eta) = \int_0^{s_0} g(s)\exp[i\eta s + i\frac{s^3}{3}]ds,$$

$$(3.16b) \qquad I_2(\eta) = \frac{i}{\eta+s_0^2}\left(g(s_0) - \frac{2is_0}{(\eta+s_0^2)^2}g(s_0) + \frac{i}{\eta+s_0^2}g^{(1)}(s_0)\right)\exp[i\eta s_0 + i\frac{s_0^3}{3}],$$

$$(3.16c) \qquad I_3(\eta) = -\int_{s_0}^{\infty} G(\eta,s)\exp[i\eta s + i\frac{s^3}{3}]ds, \quad \text{where } s_0 \equiv s_0 \text{ and}$$

$$(3.17) \qquad G(\eta,s) = \left(\frac{-2}{(\eta+s^2)^3} + \frac{12s^2}{(\eta+s^2)^4}\right)g(s) - \frac{6s}{(\eta+s^2)^3}g^{(1)}(s) + \frac{1}{(\eta+s^2)^2}g^{(2)}(s).$$

To prove this it obviously suffices to show that

$$(3.18)\qquad \lim_{R\to\infty}\int_{s_0}^{R} g(s)\exp[i\eta s + i\,\frac{s^3}{3}]ds = I_2(\eta) + I_3(\eta).$$

Let us write

$$(3.19a)\qquad \phi = \eta s + \frac{s^3}{3}, \qquad\qquad (3.19b)\qquad \psi = [\frac{\partial\phi}{\partial s}]^{-1}.$$

Then for $s \geq s_0$, $\eta \geq -\eta_0$ one has

$$(3.20)\qquad 0 < \psi = \frac{1}{\eta + s^2} \leq \frac{1}{-\eta_0 + s^2} \leq 1.$$

Now, integrating by parts twice we find for $R > s_0$

$$(3.21)\qquad \int_{s_0}^{R} e^{i\phi} g ds = -i\psi e^{i\phi} \sum_{\ell=0}^{1} (iT)^{\ell} g\Big|_{s_0}^{R} + \int_{s_0}^{R} e^{i\phi}(iT)^2 g ds,$$

where the operator T is defined by

$$(3.22)\qquad Tg = \frac{\partial}{\partial s}(\psi g).$$

Clearly,

$$(3.23a)\qquad (Tg)(\eta,s) = \frac{-2s}{(\eta + s^2)^2}\, g(s) + \frac{1}{\eta + s^2}\, g^{(1)}(s)$$

$$(3.23b)\qquad (T^2 g)(\eta,s) = G(\eta,s)$$

with G as in (3.17).

Substituting (3.23) into (3.21), taking $R \to \infty$ and using (3.10) we arrive at the desired identity (3.18), where the integral defining I_3 is absolutely convergent.

Herewith $I(\eta)$ is well-defined for $\eta \geq -\eta_0$. Next, let us show that $I \in C^2\ (-\eta_0 \leq \eta < +\infty)$ with (3.12).

Note first that I_1 and I_2 behave perfectly. Actually, both I_1 and I_2 belong to $C^{\infty}\ (-\eta_0 \leq \eta < +\infty)$ and all derivatives vanish for $\eta \to +\infty$. As for I_1, the first statement follows from dominated convergence, the second from the Riemann-Lebesgue lemma. As for I_2, both statements are evident from its explicit form (3.16b).

Thus it remains to consider I_3. From (3.17) we readily obtain that for each fixed $s \geq s_0$ the function $\eta \mapsto e^{i\phi}G(\eta,s)$ belongs to $C^2\ (-\eta_0 \leq \eta < +\infty)$ with

$$(3.24)\qquad \lim_{\eta\to+\infty} (\frac{\partial}{\partial\eta})^j (e^{i\phi}G(\eta,s)) = 0, \quad j = 0,1,2.$$

Moreover, in view of (3.10-20) there are constants $c_j(\eta_0,g)$, depending only on η_0 and g, such that for $\eta \geqq -\eta_0$, $s \geqq s_0$

(3.25) $$\left|\left(\frac{\partial}{\partial\eta}\right)^j (e^{i\phi}G(\eta,s))\right| \leqq c_j(\eta_0,g)s^{-4+j}, \quad j = 0,1,2.$$

Since the right hand side of (3.25) clearly belongs to L^1 $(s_0 \leqq s < +\infty)$, we may apply the dominated convergence theorem and conclude that $I_3 \in C^2$ $(-\eta_0 \leqq \eta < +\infty)$ with

(3.26) $$\lim_{\eta\to+\infty} \left(\frac{d}{d\eta}\right)^j I_3(\eta) = 0, \quad j = 0,1,2.$$

As a consequence $I \in C^2$ $(-\eta_0 \leqq \eta < +\infty)$ with (3.12), as was to be proven.

Finally let us prove (3.14) for $\eta \geqq -\eta_0$ under the assumption (3.13). By virtue of the dominated convergence theorem it is sufficient to show that

(3.27) $$\frac{\partial}{\partial\eta} \lim_{R\to\infty} \int_{s_0}^{R} e^{i\phi}g ds = \lim_{R\to\infty} \frac{\partial}{\partial\eta} \int_{s_0}^{R} e^{i\phi}g ds.$$

Now, insert (3.21) into (3.27). Then, by the above it is clear that (3.27) holds provided that

(3.28) $$\lim_{s\to+\infty} \frac{\partial}{\partial\eta} [-i\psi e^{i\phi} \sum_{\ell=0}^{1} (iT)^\ell g] = 0.$$

It is an easy matter to show that under the assumption (3.13) condition (3.28) is indeed satisfied.

To conclude with note that since η_0 was arbitrarily chosen we have in fact shown that $I \in C^2$ $(-\infty < \eta < +\infty)$. Likewise (3.14) under the assumption (3.13) holds for all $\eta \in \mathbb{R}$. Herewith the proof of the lemma is completed. □

Let us observe at this point that as a consequence of (3.14) and the above lemma the first derivative of the Airy function can be represented as follows

(3.29) $$Ai^{(1)}(\eta) = \frac{1}{2\pi} \lim_{R\to\infty} \int_{-R}^{R} is \exp[i\eta s + i\frac{s^3}{3}]ds, \quad \eta \in \mathbb{R}.$$

We proceed with a rather general lemma, which provides a good insight in the structure of integrals of type (3.9a) when considered in the parameter region $\varepsilon > 0$, $\eta \geqq -\nu$. To prove it we use some fruitful ideas already developed in the proof of lemma 3.1.

Lemma 3.2. *Let* b *be any function satisfying*

(3.30a) $\quad b$ *is of class* $C^2(\mathbb{R})$, *such that* $b^*(k) = b(-k)$

(3.30b) $\quad$ *The derivatives* $b^{(j)}(k)$, $j = 0,1,2$ *are bounded on* $\mathbb{R}$.

Then

(3.31) $$J(\varepsilon,\eta) = \frac{1}{2\pi}\lim_{R\to\infty}\int_{-R}^{R} b(\varepsilon s)\exp[i\eta s + i\frac{s^3}{3}]ds$$

is well-defined for all $\eta \in \mathbb{R}$, $\varepsilon > 0$ *and belongs to* C^2 $(-\infty < \eta < +\infty)$ *with*

(3.32) $$\lim_{\eta\to+\infty}\left(\frac{\partial}{\partial\eta}\right)^j J(\varepsilon,\eta) = 0, \quad j = 0,1,2.$$

Furthermore one has the representation

(3.33) $$J(\varepsilon,\eta) = b(0)Ai(\eta) - i\varepsilon b^{(1)}(0)Ai^{(1)}(\eta) + R_2(\varepsilon,\eta),$$

where the remainder term can be estimated as follows.
Let $\varepsilon > 0$ *be arbitrarily fixed.*
Then for $\eta > 0$

(3.34a) $$|R_2(\varepsilon,\eta)| \leq \varepsilon^2 \|b^{(2)}\|_\infty \frac{7}{8}\eta^{-3/2}, \quad \|b^{(2)}\|_\infty = \sup_{k\in\mathbb{R}} |b^{(2)}(k)|$$

and for $\eta \geq -\eta_0$ *with* $\eta_0 \geq 0$ *any constant*

(3.34b) $$|R_2(\varepsilon,\eta)| \leq \varepsilon^2 \|b^{(2)}\|_\infty C(\eta_0),$$

where $C(\eta_0)$ *is a constant depending only on* η_0, *which can be given in explicit form.*
If, instead of (3.30b), we make the stronger assumption

(3.35) $$b^{(j)}(k) = O(|k|^{-1}), \quad k \to \pm\infty, \quad j = 0,1,2,$$

then we also have the representation

(3.36) $$\frac{\partial}{\partial\eta}J(\varepsilon,\eta) = b(0)Ai^{(1)}(\eta) + r_1(\varepsilon,\eta), \text{ with}$$

(3.37a) $$|r_1(\varepsilon,\eta)| \leq \varepsilon\|(kb)^{(2)}\|_\infty \frac{7}{8}\eta^{-3/2}, \qquad \text{for } \varepsilon > 0,\ \eta > 0$$

(3.37b) $$|r_1(\varepsilon,\eta)| \leq \varepsilon\|(kb)^{(2)}\|_\infty C(\eta_0), \qquad \text{for } \varepsilon > 0,\ \eta \geq -\eta_0,$$

where $C(\eta_0)$ *denotes the same constant as in (3.34b).*

Remark:

(i) Clearly the condition $b^*(k) = b(-k)$ can be omitted. We include it because it simplifies the proof somewhat and $b_r(k)$ has this property.

(ii) Observe that (3.33) corresponds to the second order Taylor expansion for η fixed of the function $\varepsilon \mapsto J(\varepsilon,\eta)$ near $\varepsilon = 0$. Let us motivate this choice.

First of all, in view of (3.9) we are interested in the behaviour of $J(\varepsilon,\eta)$ as $\varepsilon \downarrow 0$. Thus it is quite natural to look at the Taylor expansion near $\varepsilon = 0$. The problem is of course: which order must we take? In the light of future estimates the answer is rather simple: such an order n that the remainder term $R_n(\varepsilon,\eta)$ is integrable over $(1 < \eta < +\infty)$. Now, with due perseverence one can show that $R_n(\varepsilon,\eta)$ behaves as $\eta^{-n+\frac{1}{2}}$ as $\eta \to +\infty$ for $\varepsilon > 0$ fixed, provided b has n bounded derivatives. Thus we must take $n \geq 2$. However, it is easily seen that choosing $n > 2$ does not improve the estimates much (in fact it does so only, when for some m the subsequent derivatives $b^{(j)}(0)$, $j = 0,1,2,\dots,m$ vanish). Hence the choice $n = 2$ is singled out.

Proof: The relation $b^*(k) = b(-k)$ enables us to rewrite (3.31) as

$$(3.38)\qquad J(\varepsilon,\eta) = \frac{1}{\pi} \lim_{R\to\infty} \mathrm{Re} \int_0^R b(\varepsilon s)e^{i\phi}ds, \qquad \phi = \eta s + \frac{s^3}{3}.$$

Hence, it is a direct consequence of lemma 3.1 that for any $\varepsilon > 0$ the function $J(\varepsilon,\eta)$ is a well-defined member of C^2 $(-\infty < \eta < +\infty)$ satisfying (3.32).

Let us prove the representation (3.33-34).
To this end we insert into (3.31) the Taylor expansion

$$(3.39a)\qquad b(k) = b(0) + kb^{(1)}(0) + b_2(k)$$

$$(3.39b)\qquad b_2(k) = \int_0^k (k - \tilde{k})b^{(2)}(\tilde{k})d\tilde{k}.$$

In view of (3.3-29) this yields (3.33) with

(3.40a) $\quad R_2(\varepsilon,\eta) = \frac{1}{2\pi} \lim_{R\to\infty} \int_{-R}^{R} b_2(\varepsilon s)e^{i\phi}ds = \frac{1}{\pi} \operatorname{Re} \tilde{R}_2(\varepsilon,\eta)$

(3.40b) $\quad \tilde{R}_2(\varepsilon,\eta) = \lim_{R\to\infty} \int_0^R b_2(\varepsilon s)e^{i\phi}ds.$

Next, fix $\varepsilon > 0$. Set $g(s) = b_2(\varepsilon s)$. Then for $s \in \mathbb{R}$

(3.41a) $\quad |g(s)| \leq \frac{1}{2}s^2\varepsilon^2\|b^{(2)}\|_\infty$

(3.41b) $\quad |g^{(1)}(s)| \leq |s|\varepsilon^2\|b^{(2)}\|_\infty$

(3.41c) $\quad |g^{(2)}(s)| \leq \varepsilon^2\|b^{(2)}\|_\infty.$

Moreover, (3.30) implies that g satisfies condition (3.10) of lemma 3.1.

To prove (3.34b) let $\eta_0 \geq 0$ be a constant. Put $s_0 = \sqrt{1+\eta_0}$. Then for $\eta \geq -\eta_0$, $\tilde{R}_2(\varepsilon,\eta)$ has the representation (3.15-16-17). Using (3.41a) we find from (3.16a)

(3.42a) $\quad |I_1| \leq \int_0^{s_0} \frac{1}{2}s^2\varepsilon^2\|b^{(2)}\|_\infty ds = \varepsilon^2\|b^{(2)}\|_\infty \frac{1}{6} s_0^3.$

Furthermore, applying (3.41a-b) and (3.20) we obtain from (3.16b)

(3.42b) $\quad |I_2| \leq \varepsilon^2\|b^{(2)}\|_\infty (s_0 + \frac{1}{2}s_0^2 + s_0^3).$

Next, combining (3.41) and (3.20) we get from (3.16c-17)

(3.42c) $\quad |I_3| \leq \varepsilon^2\|b^{(2)}\|_\infty\gamma(s_0)$ with

(3.43) $\quad \gamma(s_0) = \int_{s_0}^{\infty} \left(\frac{6s^4}{(-\eta_0+s^2)^4} + \frac{7s^2}{(-\eta_0+s^2)^3} + \frac{1}{(-\eta_0+s^2)^2}\right) ds.$

Together (3.15), (3.40) and (3.42) imply that for $\eta \geq -\eta_0$ the estimate (3.34b) holds with

(3.44) $\quad C(\eta_0) = (s_0 + \frac{1}{2}s_0^2 + \frac{7}{6}s_0^3 + \gamma(s_0))/\pi.$

To proceed with, let us prove (3.34a).

Let $\eta > 0$. Then integrating by parts twice we arrive at (3.21) with s_0 replaced by $-R$ and $g(s) = b_2(\varepsilon s)$. Taking $R \to \infty$ and exploiting (3.30) we obtain

(3.45) $\quad 2\pi R_2(\varepsilon,\eta) = -\int_{-\infty}^{\infty} G(\eta,s)e^{i\phi}ds$

with $G(\eta,s)$ given by (3.17). An application of (3.41) now gives us

$$(3.46)\qquad |R_2(\varepsilon,\eta)| \leq \frac{1}{2\pi}\varepsilon^2 \|b^{(2)}\|_\infty \int_{-\infty}^{\infty}\left(\frac{6s^4}{(\eta+s^2)^4}+\frac{7s^2}{(\eta+s^2)^3}+\frac{1}{(\eta+s^2)^2}\right)ds =$$
$$= \varepsilon^2\|b^{(2)}\|_\infty \frac{7}{8}\eta^{-3/2},$$

which proves (3.34a).

Lastly, let us show that under the condition (3.35) the representation (3.36-37) is valid. To do so, observe that for fixed $\varepsilon > 0$ the function $g(s) = b(\varepsilon s)$ easily meets the requirement (3.13) of lemma 3.1. Combining (3.14) with (3.38) we therefore have

$$(3.47a)\qquad \frac{\partial J}{\partial\eta}(\varepsilon,\eta) = \varepsilon^{-1}\tilde{J}(\varepsilon,\eta)$$

$$(3.47b)\qquad \tilde{J}(\varepsilon,\eta) = \frac{1}{2\pi}\lim_{R\to\infty}\int_{-R}^{R}\tilde{b}(\varepsilon s)e^{i\phi}ds$$

$$(3.47c)\qquad \tilde{b}(k) = ikb(k).$$

Evidently $\tilde{b}$ satisfies (3.30). Hence $\tilde{J}(\varepsilon,\eta)$ has the representation (3.33-34) with $b(k)$ replaced by $\tilde{b}(k)$. Substituting back we immediately find (3.36-37), which completes the proof of this lemma. □

Now is the time to apply the above results to the integral (3.9a), to translate from $J(\varepsilon,\eta)$-language into $\Omega_c(x+y;t)$-language and to see what we have got. Doing so we arrive at

<u>Theorem 3.3</u>. *Assume that the right reflection coefficient* $b_r(k)$ *satisfies*

(3.48) b_r *is of class* $C^2(\mathbb{R})$ *and the derivatives* $b_r^{(j)}(k)$, $j = 0,1,2$ *are bounded on* $\mathbb{R}$.

Then Ω_c *is strongly differentiable in* V *with respect to* x *at every point* (x,t), $x \in \mathbb{R}$, $t > 0$. *Let the derivative be denoted by* Ω_c'.

Furthermore, let $y > 0$, $x \in \mathbb{R}$, $t > 0$.
Let μ *and* ν *denote arbitrary nonnegative constants.*
Put

$$(3.49a)\qquad w = x + y + \mu,\quad b(k,\mu) = b_r(k)e^{-2ik\mu},\quad b^{(j)} = \left(\frac{\partial}{\partial k}\right)^j b$$

(3.49b) $\quad T = (3t)^{1/3}, \quad Z = w(3t)^{-1/3}.$

Then one has the representation

(3.50) $$\Omega_c(x+y;t) = T^{-1}b_r(0)\mathrm{Ai}(Z) - \tfrac{1}{2}iT^{-2}b^{(1)}(0,\mu)\mathrm{Ai}^{(1)}(Z) + R(Z,T,\mu)$$

with

(3.51a) $$|R(Z,T,\mu)| \leq T^{-3}\|b^{(2)}\|_\infty \frac{7}{32} Z^{-3/2} \quad \textit{for } T > 0,\ Z > 0$$

(3.51b) $$|R(Z,T,\mu)| \leq T^{-3}\|b^{(2)}\|_\infty \frac{1}{4} C(\nu) \quad \textit{for } T > 0,\ Z \geq -\nu$$

where $C(\nu)$ *denotes the constant (3.44) with* η_0 *replaced by* ν.
If we make the additional assumption

(3.52) $$b_r^{(j)}(k) = O(|k|^{-1}), \quad k \to \pm\infty, \quad j = 0,1,2,$$

then we also have

(3.53) $$\Omega_c'(x+y;t) = T^{-2}b_r(0)\mathrm{Ai}^{(1)}(Z) + r(Z,T,\mu) \quad \textit{with}$$

(3.54a) $$|r(Z,T,\mu)| \leq T^{-3}\|(kb)^{(2)}\|_\infty \frac{7}{16} Z^{-3/2} \quad \textit{for } T > 0,\ Z > 0$$

(3.54b) $$|r(Z,T,\mu)| \leq T^{-3}\|(kb)^{(2)}\|_\infty \frac{1}{2} C(\nu) \quad \textit{for } T > 0,\ Z \geq -\nu$$

with $C(\nu)$ *as in (3.51b).*

Remark. If (3.52) holds then it follows from (3.51-54) that the functions $R(Z,T,\mu)$ and $r(Z,T,\mu)$ can be estimated simultaneously by

(3.55) $$\max(|R(Z,T,\mu)|, |r(Z,T,\mu)|) \leq \rho T^{-3}(1+\nu+Z)^{-3/2} \quad \text{for } T > 0,\ Z \geq -\nu,$$

where the constant ρ is given by

(3.56a) $$\rho = DN_2$$

(3.56b) $$D = D(\nu) = \frac{7}{16}\left[1 + (1+\nu)\left(\frac{8}{7}C(\nu)\right)^{2/3}\right]^{3/2}$$

(3.56c) $$N_2 = N_2(\mu,b_r) = \max\left(\tfrac{1}{2}\|b^{(2)}\|_\infty,\ \|(kb)^{(2)}\|_\infty\right).$$

Proof: Let $\mu \geq 0$ be arbitrarily fixed. Then (3.48) obviously implies that $b(k) = b(k,\mu)$ satisfies condition (3.30) of lemma 3.2.

Now, let $J(\varepsilon,\eta)$ for $\eta \in \mathbb{R}$, $\varepsilon > 0$ be defined by (3.31). Reasoning as in the beginning of this section we then obtain for $y > 0$, $x \in \mathbb{R}$, $t > 0$

(3.57a) $\quad \Omega_c(x+y;t) = 2\varepsilon J(\varepsilon,\eta)$

(3.57b) $\quad \varepsilon = \frac{1}{2T}, \quad \eta = Z$

with T and Z as in (3.49).
The assertions of this theorem are now easily proven by combining (3.57) with lemma 3.2.
Specifically, the representation (3.50-51) is merely a transcription of (3.33-34). Furthermore, lemma 3.2 tells us that $\Omega_c(\xi;t)$ belongs to C^2 $(-\infty < \xi < +\infty)$ with

(3.58) $$\lim_{\xi\to+\infty} \left(\frac{\partial}{\partial\xi}\right)^j \Omega_c(\xi;t) = 0, \quad j = 0,1,2.$$

Plainly, this implies that for all $a \in \mathbb{R}$

(3.59) $$N(a,t) \equiv \sup_{\xi\geq a} \left|\left(\frac{\partial}{\partial\xi}\right)^2 \Omega_c(\xi;t)\right| < +\infty.$$

With the help of (3.59) it is not hard to show that Ω_c is strongly x-differentiable. Indeed, let (x,t), $x \in \mathbb{R}$, $t > 0$ be an arbitrary but fixed point. Then one has for $h \in \mathbb{R}$, $0 < |h| < 1$

(3.60) $$\sup_{0<y<+\infty} \left| \frac{\Omega_c(x+h+y;t) - \Omega_c(x+y;t)}{h} - \frac{\partial\Omega_c}{\partial x}(x+y;t)\right| =$$

$$= \sup_{0<y<+\infty} \left| \frac{1}{h} \int_0^h (h - \tilde{h}) \frac{\partial^2\Omega_c}{\partial x^2}(x + \tilde{h} + y;t)d\tilde{h}\right|$$

$$\leq \tfrac{1}{2}N(x-1,t)|h| = O(|h|) \quad \text{as } h \to 0.$$

Consequently, Ω_c is strongly x-differentiable at (x,t) with derivative

(3.61) $$\Omega_c'(x+y;t) = 4\varepsilon^2 \frac{\partial J}{\partial\eta}(\varepsilon,\eta).$$

Finally, if b_r satisfies (3.52), then b fulfills condition (3.35) of lemma 3.2. Together (3.36-37), (3.61) and (3.57b) yield the desired representation (3.53-54) and so the proof is done. □

The results laid down in theorem 3.3 are remarkable for two reasons. Firstly, they display in a strikingly explicit way the structure as well as the magnitude of the functions $y \mapsto \Omega_c(x+y;t)$, $\Omega_c'(x+y;t)$ in the parameter region $x \geq -\mu - \nu(3t)^{1/3}$. This will be extensively used below, when we start estimating in the norms (3.65). Secondly, the conditions imposed on the right reflection coefficient $b_r(k)$ are only very weak. In this respect, let us note that since any initial function $u_0(x)$ considered in this paper is supposed to satisfy at least the conditions stated at the beginning of subsection 2.1, we already know from that subsection that $b_r \in L^1 \cap C_0(\mathbb{R})$ such that $b_r(k) = o(|k|^{-1})$ for $k \to \pm\infty$. Hence, in view of an interpolation argument, the only extra condition imposed on b_r by (3.48-52) is

(3.62) $\quad b_r \in C^2(\mathbb{R})$ such that $b_r^{(2)}(k) = O(|k|^{-1})$ for $k \to \pm\infty$.

In general, initial functions satisfying (1.2) will fulfill condition (3.62). A mild algebraic decay of $u_0(x)$ and a number of its derivatives is already sufficient. Specifically if u_0 satisfies (1.3) then b_r belongs at least to $C^6(\mathbb{R})$ with $b_r^{(j)}(k) = O(|k|^{-5})$, $k \to \pm\infty$, $j = 0,1,2,\ldots,6$, so that (3.62) is amply fulfilled.

We now turn to the construction of bounds for Ω_c, Ω_c' in various settings. To avoid any misunderstandings and to have an easy reference we stress the following:

From now on we shall assume that the reflection coefficient $b_r(k)$ fulfills the requirements of theorem 3.3, i.e.:

(3.63) $\quad b_r$ is of class $C^2(\mathbb{R})$ and the derivatives $b_r^{(j)}(k)$, $j = 0,1,2$ satisfy

$$b_r^{(j)}(k) = O(|k|^{-1}), \quad k \to \pm\infty.$$

With the help of theorem 3.3 it is now an easy matter to show that in the parameter region

(3.64) $\quad T = (3t)^{1/3} \geq 1, \quad x \geq -\mu - \nu T$, where μ and ν are nonnegative constants, one has the estimates

$$\|\Omega_c\| = \sup_{0<y<+\infty} |\Omega_c(x+y;t)| \leq \gamma T^{-1}, \tag{3.65a}$$

$$\|\Omega_c\|_{L^1} = \int_0^\infty |\Omega_c(x+y;t)|dy \leq |b_r(0)|\left(\frac{1}{3}+ \int_{-\nu}^{0} |Ai(\eta)|d\eta\right) + \gamma T^{-1}, \tag{3.65b}$$

$$\|\Omega_c'\| = \sup_{0<y<+\infty} |\Omega_c'(x+y;t)| \leq \gamma T^{-2}, \tag{3.65c}$$

$$\|\Omega_c'\|_{L^1} = \int_0^\infty |\Omega_c'(x+y;t)|dy \leq \gamma T^{-1}, \tag{3.65d}$$

where the constant γ is given by

$$\gamma = AN_1 + BN_2, \text{ with } N_2 \text{ as in (3.56c) and} \tag{3.66a}$$

$$A = A(\nu) = \max\left(\sup_{\eta\geq-\nu} |Ai(\eta)|, \sup_{\eta\geq-\nu} |Ai^{(1)}(\eta)|, \int_{-\nu}^{\infty} |Ai^{(1)}(\eta)|d\eta\right), \tag{3.66b}$$

$$B = B(\nu) = \tfrac{1}{2}C(\nu)(1+\nu) + \frac{7}{8} \text{ with } C(\nu) \text{ as in (3.51b)}, \tag{3.66c}$$

$$N_1 = N_1(\mu,b_r) = \max(2|b_r(0)|, |b^{(1)}(0,\mu)|) \text{ with } \tag{3.66d}$$

$$b(k,\mu) = b_r(k)e^{-2ik\mu}.$$

Note that (3.65b) hinges on property (3.8) of the Airy function.

Next, let us apply the above results to investigate the mapping T_c in the parameter region (3.64). By theorem 3.3 and (3.7-55) there is in this region a function $c(t)$ such that

$$|\Omega_c(x+y;t)| + |\Omega_c'(x+y;t)| \leq c(t)(1+y)^{-3/2}. \tag{3.67}$$

As a result, the function

$$(T_c g)(y) = \int_0^\infty \Omega_c(x+y+z;t)g(z)dz \tag{3.68}$$

is continuous in y, since the integrand is dominated by $c(t)(1+z)^{-3/2}\|g\|$. On the other hand

$$\|T_c g\| \leq \|g\| \int_0^\infty |\Omega_c(x+y;t)|dy. \tag{3.69}$$

Hence, in view of (3.65b), T_c is a continuous mapping of V into V with a norm that satisfies

$$(3.70)\qquad \|T_c\| \le |b_r(0)|\left(\frac{1}{3} + \int_{-\nu}^{0} |Ai(\eta)|d\eta\right) + \gamma T^{-1}.$$

Finally, reasoning as in Chapter 1, we extract from (3.67), that T_c is strongly x-differentiable in V with derivative

$$(3.71)\qquad (T_c'g)(y) = \int_0^{\infty} \Omega_c'(x+y+z;t)g(z)dz.$$

From (3.65d) we find in the region (3.64) the following estimate

$$(3.72)\qquad \|T_c'\| \le \gamma T^{-1}.$$

For future reference, let us note that, in addition to (3.65), theorem 3.3 gives us useful bounds containing both x and t. In particular, fixing $\theta \in (0,1)$ in (3.7), we obtain in the region (3.64)

$$(3.73)\qquad \max\left[T^{-1}\|\Omega_c\|,\ \|\Omega_c'\|\right] \le \gamma_\theta T^{-2}\exp\left[-\frac{2}{3}\theta\left(\nu + \frac{x+\mu}{T}\right)^{3/2}\right] + \rho T^{-3}\left(1 + \nu + \frac{x+\mu}{T}\right)^{-3/2},$$

where the constant γ_θ is given by

$$(3.74)\qquad \gamma_\theta = \tilde{A}N_1,\quad \tilde{A} = \tilde{A}(\theta,\nu) = \max(a_{0,\theta}(\nu),\ a_{1,\theta}(\nu))$$

with the constants ρ, N_1, $a_{m,\theta}(\nu)$ as in (3.56a-66d-7) respectively.

Of course, in the degenerate exceptional case (2.20) the estimates (3.65-70-72-73) can be improved. Specifically, then theorem 3.3 tells us that in the parameter region (3.64)

$$(3.75a)\qquad \|\Omega_c\| \le \gamma T^{-2} \qquad\qquad \|T_c\| \le \gamma T^{-1}$$

$$(3.75b)\qquad \|\Omega_c'\| \le \gamma T^{-3} \qquad\qquad \|T_c'\| \le \gamma T^{-2},$$

with γ still given by (3.66a). Moreover, (3.73) holds with the factor T^{-2} in front of the exponential function on the right replaced by T^{-3}. It is easily verified that further simplifications of this type occur when also $b_r^{(1)}(0)$ vanishes.

As mentioned in subsection 2.1 it rarely occurs that $b_r(0) = 0$. However, in the usual case $b_r(0) \neq 0$ one can still simplify the discussion somewhat by working in the specially selected parameter region (3.64) with μ chosen to be $\mu = b_r^{(1)}(0)/(2ib_r(0))$, in which case the derivative of the Airy function disappears from (3.50). In the present discussion however we shall stick to (3.64) with μ arbitrary.

4. Solution of the KdV initial value problem in the absence of solitons.

In this section we study the asymptotic behaviour of the solution $u(x,t)$ of the KdV equation evolving from an initial function $u_0(x)$, which generates no bound states in the Schrödinger scattering problem. It is assumed that the condition (3.63) is fulfilled.

We shall work in the coordinate region $t \geq t_c$, $x \geq -\zeta$, $\zeta = \mu + \nu T$, $T = (3t)^{1/3}$, where μ, ν and t_c are nonnegative constants, with ν and $t_c = t_c(\mu,\nu,b_r)$ to be specified presently.

Clearly, in the absence of bound states the Gel'fand-Levitan equation reduces to

(4.1) $$(I + T_c)\beta = -\Omega_c.$$

For notational convenience we introduce the number $\nu_c > 0$, uniquely determined by

(4.2) $$\int_{-\nu_c}^{0} |Ai(\eta)|d\eta = \frac{2}{3}.$$

The existence of ν_c is guaranteed by the fact (see [14]) that $Ai(\eta)$ is not absolutely integrable over $-\infty < \eta < 0$. As for the numerical value of ν_c, we obtain from [3], p. 478

(4.3) $$\nu_c = 1.39.$$

Moreover, [3] tells us that

(4.4) $$Ai(\eta) > 0 \text{ for } \eta \geq -\nu_c.$$

Let us now work out our specification procedure.

Firstly, we select ν such that

(4.5) $$0 \leqq \nu < \nu_c.$$

As a consequence of (4.2) and (4.4) one then has

(4.6) $$a_{0,1}(\nu) \equiv \frac{1}{3} + \int_{-\nu}^{0} |Ai(\eta)|d\eta < 1.$$

Secondly, we fix $\mu \geqq 0$ independently of ν.
Thirdly, bearing in mind that $|b_r(0)| \leqq 1$, we select t_c such that

(4.7) $$t_c > \frac{1}{3} \max \left[1, \gamma^3\left(1 - |b_r(0)|a_{0,1}(\nu)\right)^{-3}\right]$$

with $\gamma = \gamma(\mu,\nu,b_r)$ as in (3.66a).

After the above specification it is clear from (3.70) that in the parameter region $t \geqq t_c$, $x \geqq -\zeta$, $\zeta = \mu + \nu T$, $T = (3t)^{1/3}$ the operator T_c occurring in (4.1) can be estimated as follows

(4.8) $$\|T_c\| \leqq \sigma < 1 \text{ with } \sigma = |b_r(0)|a_{0,1}(\nu) + \gamma(3t_c)^{-1/3}.$$

This implies that $I + T_c$ is invertible on the Banach space V.
As a result, (4.1) has a unique solution $\beta \in V$ satisfying

(4.9) $$\|\beta\| \leqq \omega_1\|\Omega_c\|, \quad \text{with } \omega_1 = (1 - \sigma)^{-1}.$$

Furthermore, (4.1) implies that β is strongly x-differentiable with derivative

(4.10) $$\beta' = -(I + T_c)^{-1}(T_c'\beta + \Omega_c').$$

From (4.8-9-10) and (3.72) we obtain the estimate

(4.11) $$\|\beta'\| \leqq \omega_0 T^{-1}\|\Omega_c\| + \omega_1\|\Omega_c'\|, \quad \text{with } \omega_0 = \omega_1^2\gamma.$$

Recall that

(4.12) $$u(x,t) = -\frac{\partial}{\partial x}\beta(0^+;x,t).$$

Since

(4.13) $$\left|\frac{\partial}{\partial x}\beta(0^+;x,t)\right| \leqq \sup_{0<y<+\infty}\left|\frac{\partial}{\partial x}\beta(y;x,t)\right| = \|\beta'\|,$$

we arrive at the following result

(4.14) $$|u(x,t)| \leq \omega_0 T^{-1} \|\Omega_c\| + \omega_1 \|\Omega_c'\|.$$

In particular, it follows from (3.65a-c) that

(4.15) $$|u(x,t)| \leq \omega_2 T^{-2}, \text{ with } \omega_2 = (\omega_0 + \omega_1)\gamma.$$

The above results are summarized in

Theorem 4.1. *Let* $u(x,t)$ *be the solution of the Korteweg-de Vries problem*

(4.16) $$\begin{cases} u_t - 6uu_x + u_{xxx} = 0, & -\infty < x < +\infty,\ t > 0 \\ u(x,0) = u_0(x), & \end{cases}$$

where the real initial function $u_0(x)$ *is sufficiently smooth and decays sufficiently rapidly for* $|x| \to \infty$ *for the whole of the inverse scattering method to work and to guarantee the regularity and decay property (3.63) of the reflection coefficient* $b_r(k)$. *Assume that, as a potential in the Schrödinger scattering problem,* $u_0(x)$ *generates no bound states. Let* μ, ν *and* t_c *be nonnegative constants, with* ν *and* t_c *satisfying (4.5) and (4.7) respectively.*
Then, in the coordinate region $t \geq t_c$, $x \geq -\zeta$, $\zeta = \mu + \nu T$, $T = (3t)^{1/3}$ *one has the following estimate of the solution*

(4.17) $$|u(x,t)| \leq \omega_0 T^{-1} \sup_{0<y<+\infty} |\Omega_c(x+y;t)| + \omega_1 \sup_{0<y<+\infty} \left|\frac{\partial}{\partial x}\Omega_c(x+y;t)\right|,$$

where ω_0 *and* ω_1 *are the constants introduced in (4.11) and (4.9) respectively and* $\Omega_c(x+y;t)$ *is given by (3.1).*
With the constant ω_2 *as in (4.15) we have for* $t \geq t_c$

(4.18) $$\sup_{x\geq-\zeta} |u(x,t)| \leq \omega_2 T^{-2}.$$

Let us mention some implications of the preceding theorem. As a consequence of (4.17) and (3.73) the solution $u(x,t)$ of (4.16) satisfies, in the coordinate region $t \geq t_c$, $x \geq -\zeta$, $\zeta = \mu + \nu T$, $T = (3t)^{1/3}$, the following x and t dependent bound

(4.19a) $$|u(x,t)| \leq aT^{-2}\exp\left[-\frac{2}{3}\theta\left(\frac{x+\zeta}{T}\right)^{3/2}\right] + bT^{-3}\left(1 + \frac{x+\zeta}{T}\right)^{-3/2}$$

(4.19b) $a = (\omega_0 + \omega_1)\gamma_\theta, \qquad b = (\omega_0 + \omega_1)\rho,$

with $\theta \in (0,1)$, γ_θ, ρ the constants introduced at the end of section 3. Hence, for $t \geq t_c$

$$\int_{-\zeta}^{\infty} |u(x,t)|dx \leq aT^{-1} \int_0^{\infty} \exp\left[-\frac{2}{3}\theta\, s^{3/2}\right]ds + 2bT^{-2}. \tag{4.20}$$

Combining (4.18-20) with the formulae (see [18])

$$\int_{-\infty}^{\infty} u(x,t)dx = -\frac{2}{\pi}\int_0^{\infty} \log(1 - |b_r(k)|^2)dk \tag{4.21a}$$

$$\int_{-\infty}^{\infty} u^2(x,t)dx = -\frac{8}{\pi}\int_0^{\infty} k^2\log(1 - |b_r(k)|^2)dk, \tag{4.21b}$$

we obtain

$$\int_{-\infty}^{-\zeta} u(x,t)dx = -\frac{2}{\pi}\int_0^{\infty} \log(1 - |b_r(k)|^2)dk + O(t^{-1/3}) \text{ as } t \to \infty \tag{4.22a}$$

$$\int_{-\infty}^{-\zeta} u^2(x,t)dx = -\frac{8}{\pi}\int_0^{\infty} k^2\log(1 - |b_r(k)|^2)dk + O(t^{-1}) \text{ as } t \to \infty. \tag{4.22b}$$

Remark.

(i) Observe that (4.22a) can also be derived from (2.28), since by (3.65a) and (4.9)

$$\int_{-\zeta}^{\infty} u(x,t)dx = \beta(0^+;-\zeta,t) = O(t^{-1/3}) \text{ as } t \to \infty. \tag{4.23}$$

(ii) If u_0 satisfies (1.3) then all of the conditions of theorem 4.1 are fulfilled.

(iii) In the degenerate exceptional case (2.20) the above estimates can be improved. For instance, in (4.18-19) we can replace T^{-2} by T^{-3}.

(iv) For a physical interpretation of (4.21-22-23) we refer to the discussion in section 7.

5. The operator $(I + T_d)^{-1}$.

When, as a potential in the Schrödinger scattering problem, $u_0(x)$ produces bound states, then the operator $I + T_d$ makes his entrance in the Gel'fand-Levitan equation. The following lemma shows that this operator has a nice inverse.

Lemma 5.1. *For any value of the parameters* $x \in \mathbb{R}$, $t > 0$, *the operator* $I + T_d$ *is invertible on the Banach space* V *with inverse* $S = (I + T_d)^{-1}$ *given by*

$$(5.1a) \qquad (Sf)(y) = f(y) - \sum_{j=1}^{N} A_j e^{-2\kappa_j y}$$

$$(5.1b) \qquad A_j = \sum_{i=1}^{N} \beta_{ij}\left(2 \int_0^\infty e^{-2\kappa_i z} f(z)dz\right),$$

where (β_{ij}) *is the inverse of the matrix* $A = ([c_j^r(t)]^{-2} e^{2\kappa_j x} \delta_{ij} + (\kappa_i + \kappa_j)^{-1})$. *Furthermore,* S *and its strong* x*-derivative* S' *satisfy the bounds*

$$(5.2) \qquad \|S\| \leq a_0, \quad \|S'\| \leq a_1, \quad x \in \mathbb{R}, \quad t > 0,$$

$$(5.3a) \qquad a_0 = 1 + \sum_{i,j=1}^{N} \frac{1}{\kappa_i} N_{ij}, \quad a_1 = 2a_0 \sum_{i,j=1}^{N} \frac{\kappa_j}{\kappa_i} N_{ij},$$

$$(5.3b) \qquad N_{ij} = 2(\kappa_i \kappa_j)^{\frac{1}{2}} \prod_{\substack{\ell=1 \\ \ell \neq i}}^{N} \left|\frac{\kappa_i + \kappa_\ell}{\kappa_i - \kappa_\ell}\right| \prod_{\substack{p=1 \\ p \neq j}}^{N} \left|\frac{\kappa_j + \kappa_p}{\kappa_j - \kappa_p}\right| .$$

Thus, $\|S\|$ *and* $\|S'\|$ *are uniformly bounded for* $x \in \mathbb{R}$, $t > 0$ *and the bounds are explicitly given in terms of the* κ_j *only.*

Proof: For any fixed $x \in \mathbb{R}$, $t > 0$ we may solve the equation

$$(5.4) \qquad (I + T_d)g = f, \quad f,g \in V$$

to find

$$(5.5) \qquad g(y) = f(y) - \sum_{j=1}^{N} A_j e^{-2\kappa_j y},$$

where the A_j satisfy

$$(5.6a) \qquad \sum_{j=1}^{N} \alpha_{ij} A_j = 2 \int_0^\infty e^{-2\kappa_i z} f(z)dz, \quad i = 1,2,\dots,N$$

$$(5.6b) \qquad \alpha_{ij} = \alpha_j \delta_{ij} + \frac{1}{\kappa_i + \kappa_j}, \quad \alpha_j = [c_j^r(t)]^{-2} e^{2\kappa_j x}.$$

We shall show that the matrix $A = (\alpha_{ij})$ has positive determinant and thus has an inverse $A^{-1} = (\beta_{ij})$, so that $I + T_d$ is an invertible operator on the Banach space V with inverse $S = (I + T_d)^{-1}$ given by (5.1).

Furthermore, we shall prove that the matrix elements β_{ij} are uniformly bounded for $x \in \mathbb{R}$, $t > 0$, where the bound is explicitly given in terms of the κ_j by

$$(5.7)\qquad |\beta_{ij}| \le 2(\kappa_i\kappa_j)^{\frac{1}{2}} \prod_{\substack{\ell=1\\ \ell\ne i}}^{N} \left|\frac{\kappa_i+\kappa_\ell}{\kappa_i-\kappa_\ell}\right| \prod_{\substack{p=1\\ p\ne j}}^{N} \left|\frac{\kappa_j+\kappa_p}{\kappa_j-\kappa_p}\right| \equiv N_{ij}.$$

To achieve our goal, let us first introduce some notation. By $\mathcal{H}$ we denote the real Hilbert space $L^2(0,\infty)$ with inner product $\langle f,g\rangle = {}_0\!\int^\infty f(y)g(y)dy$. In $\mathcal{H}$ we consider the elements e_j defined by $e_j(y) = e^{-\kappa_j y}$. We write $\tilde{A}$ for the Gram matrix of the vectors $e_1,e_2,\ldots,e_N$, i.e. $\tilde{A} = (\tilde{\alpha}_{ij})$, $\tilde{\alpha}_{ij} = \langle e_i,e_j\rangle = (\kappa_i+\kappa_j)^{-1}$.
Since the vectors $e_1,e_2,\ldots,e_N$ are linearly independent, it is clear [5], that $\det\tilde{A} > 0$.
Let us write $(\tilde{A})^{-1} = (\tilde{\beta}_{ij})$.
Next, select $f_1,f_2,\ldots,f_N \in \mathcal{H}$ such that $\langle f_i,e_j\rangle = 0$ and $\langle f_i,f_j\rangle = \alpha_j\delta_{ij}$ for $i,j = 1,2,\ldots,N$. Put $g_j = f_j + e_j$. Then $\alpha_{ij} = \langle g_i,g_j\rangle$, which shows that $A = (\alpha_{ij})$ is the Gram matrix of the vectors $g_1,g_2,\ldots,g_N$. Since the vectors $e_1,e_2,\ldots,e_N$ are linearly independent, the same holds for $g_1,g_2,\ldots,g_N$. Hence we have proven, that $\det A > 0$ and the existence of $A^{-1} = (\beta_{ij})$ is guaranteed.

To obtain the estimate (5.7) we introduce the vectors $h_i = \sum_{j=1}^{N} \tilde{\beta}_{ij}e_j$. Clearly, $\langle h_i,e_j\rangle = \delta_{ij}$ and $\langle h_i,h_j\rangle = \tilde{\beta}_{ij}$.
Now let P be the projection of $\mathcal{H}$ onto span $\{g_1,g_2,\ldots,g_N\}$. Then

$$(5.8)\qquad Ph_i = \sum_{j,k=1}^{N} \beta_{kj}\langle h_i,g_k\rangle g_j = \sum_{j=1}^{N} \beta_{ij}g_j,$$

so that $\langle Ph_i,h_j\rangle = \beta_{ij}$ and

$$(5.9)\qquad |\beta_{ij}|^2 \le \langle h_i,h_i\rangle\langle h_j,h_j\rangle = \tilde{\beta}_{ii}\tilde{\beta}_{jj}.$$

By direct calculation (see [5]) we obtain

$$(5.10)\qquad \tilde{\beta}_{\ell\ell} = \frac{\det\left(\frac{1}{\kappa_i+\kappa_j}\right)^{N}_{i,j=1,\ i\ne\ell,\ j\ne\ell}}{\det\left(\frac{1}{\kappa_i+\kappa_j}\right)^{N}_{i,j=1}} = 2\kappa_\ell \prod_{\substack{i=1\\ i\ne\ell}}^{N} \left(\frac{\kappa_\ell+\kappa_i}{\kappa_\ell-\kappa_i}\right)^2$$

and so the proof of (5.7) is complete.
Note, that by (5.1b-7)

$$(5.11)\qquad |A_j| \le \|f\| \sum_{i=1}^{N} \frac{1}{\kappa_i} N_{ij},$$

which implies the bound (5.2-3) for $\|S\|$.

It remains to estimate the strong x-derivative of S which by (5.1) is given by

$$(5.12a)\qquad (S'f)(y) = -\sum_{j=1}^{N} A_j' e^{-2\kappa_j y}$$

$$(5.12b)\qquad A_j' = -2\kappa_j A_j + \sum_{i,p=1}^{N} \beta_{ij} \frac{2\kappa_p}{\kappa_i+\kappa_p} A_p .$$

Using (5.7-11-12b) one gets

$$(5.13)\qquad |A_j'| \leq \|f\|\left(2\kappa_j \sum_{i=1}^{N} \frac{1}{\kappa_i} N_{ij} + \sum_{i,\ell,p=1}^{N} \frac{1}{\kappa_\ell} N_{\ell p} \frac{2\kappa_p}{\kappa_i+\kappa_p} N_{ij}\right),$$

from which the bound (5.2-3) for $\|S'\|$ is an immediate consequence.

□

Remark. From the above proof it is clear that lemma 5.1 is still valid (with the same bounds (5.2-3)) if the time evolution of the normalization coefficients is not given by (2.24) - as prescribed by the KdV equation - but is instead completely arbitrary.

Corollary to lemma 5.1. *Let* $x \in \mathbb{R}$ *and* $t > 0$. *Then the equation*

$$(5.14)\qquad (I + T_d)\beta = -\Omega_d$$

admits a unique solution $\beta_d \in V$ *and we have*

$$(5.15a)\qquad \beta_d(y;x,t) = -2 \sum_{i,j=1}^{N} \beta_{ij} e^{-2\kappa_j y}$$

$$(5.15b)\qquad \beta_d'(y;x,t) = 4 \sum_{\ell,p=1}^{N} \kappa_p \beta_{\ell p} \left(e^{-2\kappa_p y} - \sum_{i,j=1}^{N} \frac{1}{\kappa_i+\kappa_p} \beta_{ij} e^{-2\kappa_j y}\right).$$

Remark. Let us recall that β_d produces the pure N-soliton solution of the KdV equation associated with $u_0(x)$ through the formula

$$(5.16a)\qquad u_d(x,t) = -\frac{\partial}{\partial x} \beta_d(0^+;x,t)$$

$$= -4 \sum_{\ell,p=1}^{N} \kappa_p \beta_{\ell p} \left(1 - \sum_{i,j=1}^{N} \frac{1}{\kappa_i+\kappa_p} \beta_{ij}\right).$$

Since

$$(5.17)\qquad 1-\sum_{i,j=1}^{N}\frac{1}{\kappa_i+\kappa_p}\beta_{ij}=[c_p^r(t)]^{-2}e^{2\kappa_p x}\sum_{\ell=1}^{N}\beta_{\ell p},$$

we find for u_d two additional representations that are useful as well:

$$(5.16b)\qquad u_d(x,t)=-4\sum_{p=1}^{N}\kappa_p[c_p^r(t)]^{-2}e^{2\kappa_p x}\left(\sum_{\ell=1}^{N}\beta_{\ell p}\right)^2,$$

$$(5.16c)\qquad u_d(x,t)=-4\sum_{p=1}^{N}\kappa_p[c_p^r(t)]^{2}e^{-2\kappa_p x}\left(1-\sum_{i,j=1}^{N}\frac{1}{\kappa_i+\kappa_p}\beta_{ij}\right)^2.$$

Consequently

$$(5.18)\qquad 0\geq u_d(x,t)\geq -4a_0\sum_{i,j=1}^{N}\kappa_i N_{ij},$$

so that $u_d(x,t)$ is uniformly bounded for $x\in\mathbb{R}$, $t>0$ and the bound does not involve the c_j^r but depends only on the κ_j in a simple explicit way. Combining (5.7-16) and (2.6-24) we obtain

$$(5.19a)\qquad \int_{-\infty}^{0}|u_d(x,t)|dx=O(e^{-8\kappa_N^3 t})\qquad \text{as } t\to\infty$$

$$(5.19b)\qquad \int_{-\infty}^{0}u_d^2(x,t)dx=O(e^{-16\kappa_N^3 t})\qquad \text{as } t\to\infty.$$

Recall that (see [10])

$$(5.20a)\qquad \int_{-\infty}^{\infty}u_d(x,t)dx=-4\sum_{p=1}^{N}\kappa_p$$

$$(5.20b)\qquad \int_{-\infty}^{\infty}u_d^2(x,t)dx=\frac{16}{3}\sum_{p=1}^{N}\kappa_p^3.$$

Starting from (5.16b-c) it is shown in [15] that as t approaches infinity the pure N-soliton solution decomposes into N solitons uniformly with respect to x on $\mathbb{R}$. More precisely one has

$$(5.21a)\qquad \lim_{t\to\infty}\sup_{x\in\mathbb{R}}\left|u_d(x,t)-\sum_{p=1}^{N}(-2\kappa_p^2\operatorname{sech}^2[\kappa_p(x-x_p^+-4\kappa_p^2t)])\right|=0,$$

$$(5.21b)\qquad x_p^+=\frac{1}{2\kappa_p}\log\left\{\frac{[c_p^r]^2}{2\kappa_p}\prod_{\ell=1}^{p-1}\left(\frac{\kappa_\ell-\kappa_p}{\kappa_\ell+\kappa_p}\right)^2\right\}.$$

6. Solution of the Gel'fand-Levitan equation in the presence of bound states.

Under the condition (3.63) we now proceed to investigate the full Gel'fand-Levitan equation

(6.1) $$(I + T_d + T_c)\beta = -\Omega$$

in the parameter region $t \geq t_{cd}$, $x \geq -\zeta$, $\zeta = \mu + \nu T$, $T = (3t)^{1/3}$, where μ, ν and t_{cd} are nonnegative constants, with ν satisfying (4.5) and with $t_{cd} = t_{cd}(\mu,\nu,b_r,\kappa_1,\kappa_2,\ldots,\kappa_N) > \frac{1}{3}$ to be specified shortly.
Applying the operator $S = (I + T_d)^{-1}$ we can rewrite (6.1) as

(6.2) $$(I + ST_c)\beta = -S\Omega.$$

To ensure the invertibility of the operator $I + ST_c$, it suffices to prove that ST_c has norm smaller than 1, as was the case in our comoving coordinate analysis [8]. However, in the present situation the operator ST_c is less manageable. Let us circumvent this difficulty and consider the operator T_cS instead.
From (5.1-11) and the estimate

(6.3) $$\left|\int_0^\infty \Omega_c(x+y+z;t)e^{-2\kappa_j z}dz\right| \leq \frac{1}{2\kappa_j}\|\Omega_c\|$$

we obtain

(6.4) $$\|T_cS\| \leq \|T_c\| + \|\Omega_c\| \sum_{i,j=1}^{N} \frac{1}{2\kappa_i\kappa_j} N_{ij}.$$

Hence, by (3.65a-70)

(6.5a) $$\|T_cS\| \leq |b_r(0)|a_{0,1}(\nu) + \tilde{\gamma}T^{-1},$$ where $a_{0,1}(\nu)$ is given by (4.6) and

(6.5b) $$\tilde{\gamma} = \gamma(1 + \sum_{i,j=1}^{N} \frac{1}{2\kappa_i\kappa_j} N_{ij}).$$

We now select t_{cd} such that

(6.6) $$t_{cd} > \frac{1}{3}\max\left[1,\tilde{\gamma}^3\left(1 - |b_r(0)|a_{0,1}(\nu)\right)^{-3}\right].$$

For $t \geq t_{cd}$ we then have

(6.7) $\|T_c S\| \leq \tilde{\sigma} < 1$, with $\tilde{\sigma} = |b_r(0)| a_{0,1}(\nu) + \tilde{\gamma}(3t_{cd})^{-1/3}$.

This shows, that the operator $I + T_c S$ is invertible on the Banach space V. Consequently, the same holds for $I + ST_c$ and we have

(6.8a) $(I + ST_c)^{-1} = I - S(I + T_c S)^{-1} T_c$,

but also

(6.8b) $(I + ST_c)^{-1} = S(I + T_c S)^{-1} S^{-1}$.

We conclude that, in the parameter region $t \geq t_{cd}$, $x \geq -\zeta$, $\zeta = \mu + \nu T$, $T = (3t)^{1/3}$, the equation (6.1) has a unique solution $\beta \in V$. This solution can be represented in terms of S, T_c and Ω by means of a Neumann series:

(6.9) $\beta = \sum_{m=0}^{\infty} (-ST_c)^m (-S\Omega)$.

Note that, while in general this series converges rather slowly, the convergence in the degenerate exceptional case (2.20) is rapid for large t by virtue of (3.75a).

7. Decomposition of the solution of the KdV problem when the initial data generate solitons.

Let us put

(7.1) $\beta = \beta_d + \beta_c$, with

(7.2) $\beta_d = -S\Omega_d$.

Introducing the decomposition (7.1) into (6.2), we find

(7.3) $(I + ST_c)\beta_c = -S(\Omega_c + T_c \beta_d)$.

From (6.8b) it is clear that (7.3) has a unique solution $\beta_c \in V$, satisfying

(7.4) $\qquad \|\beta_c\| \leq (1 - \|T_c S\|)^{-1}\|S\|\left(\|\Omega_c\| + \|T_c\beta_d\|\right).$

By (5.7-15a) and (6.3) one has

(7.5) $\qquad \|T_c\beta_d\| \leq \|\Omega_c\| \sum_{i,j=1}^{N} \frac{1}{\kappa_i} N_{ij}$

so that, in view of (5.2-3a) and (6.7),

(7.6a) $\qquad \|\beta_c\| \leq \tilde{\omega}_1\|\Omega_c\|$, with

(7.6b) $\qquad \tilde{\omega}_1 = (1 - \tilde{\sigma})^{-1} a_0^2.$

Let us keep in mind that the solution of the KdV equation is given by

(7.7) $\qquad u(x,t) = u_d(x,t) - \frac{\partial}{\partial x}\beta_c(0^+;x,t),$

where $u_d(x,t)$ denotes the pure N-soliton solution introduced in (5.16). Therefore, we need estimates of the derivative of β_c with respect to x. From (6.8) and (7.3) it is clear that β_c is strongly x-differentiable, the derivative β_c' being uniquely determined by

(7.8) $\qquad (I + ST_c)\beta_c' = -S\{T_c'(\beta_c+\beta_d) + \Omega_c' + T_c\beta_d'\} - S'\{\Omega_c + T_c(\beta_c+\beta_d)\}.$

Using (5.3a-7-15b) and (6.3) one gets

(7.9) $\qquad \|T_c\beta_d'\| \leq 2a_0\|\Omega_c\| \sum_{i,j=1}^{N} N_{ij}.$

Furthermore, (3.71) and (5.7-15a) imply

(7.10) $\qquad \|T_c'\beta_d\| \leq \|\Omega_c'\| \sum_{i,j=1}^{N} \frac{1}{\kappa_i} N_{ij}.$

Combining (3.72), (5.2-3a), (6.7-8) and (7.5-6-9-10) we obtain from (7.8) the following estimate

(7.11a) $\qquad \|\beta_c'\| \leq \tilde{\omega}_0\|\Omega_c\| + \tilde{\omega}_1\|\Omega_c'\|$, with

(7.11b) $\qquad \tilde{\omega}_0 = \tilde{\omega}_1 a_0^{-1}\left(\tilde{\omega}_1\tilde{\sigma} + 2a_0 \sum_{i,j=1}^{N} N_{ij}\right) + a_1 a_0^{-1}(a_0 + \tilde{\omega}_1\tilde{\sigma})^2.$

Evidently

(7.12) $\qquad \left| -\frac{\partial}{\partial x}\beta_c(0^+;x,t)\right| \leq \|\beta_c'\|.$

By virtue of (3.65a-c) this yields

(7.13) $$|-\frac{\partial}{\partial x}\beta_c(0^+;x,t)| \le \tilde{\omega}_2 T^{-1}, \quad \text{with } \tilde{\omega}_2 = (\tilde{\omega}_0 + \tilde{\omega}_1)\gamma.$$

Summarizing the above results we obtain

Theorem 7.1. *Let* $u(x,t)$ *be the solution of the Korteweg-de Vries problem*

(7.14) $$\begin{cases} u_t - 6uu_x + u_{xxx} = 0, & -\infty < x < +\infty, \quad t > 0 \\ u(x,0) = u_0(x) \end{cases}$$

where the real initial function $u_0(x)$ *is sufficiently smooth and decays sufficiently rapidly for* $|x| \to \infty$ *for the whole of the inverse scattering method to work and to guarantee the regularity and decay property (3.63) of the reflection coefficient* $b_r(k)$. *Assume that, as a potential in the Schrödinger scattering problem,* $u_0(x)$ *produces* $N \ge 1$ *bound states.*
Let μ, ν *and* t_{cd} *be nonnegative constants, with* ν *and* t_{cd} *satisfying (4.5) and (6.6) respectively.*
Then, in the coordinate region $t \ge t_{cd}$, $x \ge -\zeta$, $\zeta = \mu + \nu T$, $T = (3t)^{1/3}$ *one has the following decomposition of the solution*

(7.15a) $$u(x,t) = u_d(x,t) + u_c(x,t),$$

(7.15b) $$|u_c(x,t)| \le \tilde{\omega}_0 \sup_{0<y<+\infty} |\Omega_c(x+y;t)| + \tilde{\omega}_1 \sup_{0<y<+\infty} |\frac{\partial}{\partial x}\Omega_c(x+y;t)|,$$

where $u_d(x,t)$ *is the pure N-soliton solution (5.16),* $\tilde{\omega}_0$ *and* $\tilde{\omega}_1$ *are the constants introduced in (7.11b) and (7.6b) respectively and* $\Omega_c(x+y;t)$ *is given by (3.1).*
With the constant $\tilde{\omega}_2$ *as in (7.13) we have for* $t \ge t_{cd}$

(7.16) $$\sup_{x \ge -\zeta} |u_c(x,t)| \le \tilde{\omega}_2 T^{-1}.$$

Evidently, in the degenerate exceptional case (2.20) the estimate (7.16) can be improved, since T^{-1} can be replaced by T^{-2}. Although similar remarks apply to the estimates below, they are omitted to avoid interference with the reasoning.

Let us emphasize that all of the requirements of theorem 7.1 are fulfilled if u_0 satisfies (1.3).

Theorem 7.1 has a number of interesting consequences.

Firstly, by combining (7.16) with (5.21) it is found that the solution $u(x,t)$ of (7.14) splits up into N solitons as $t \to \infty$ in the following way:

Corollary to theorem 7.1. *Let* $\{b_r(k), \kappa_1 > \kappa_2 > \ldots > \kappa_N, c_1^r, c_2^r, \ldots, c_N^r\}$ *be the right scattering data associated with* $u_0(x)$. *Then the solution of (7.14) satisfies*

$$\text{(7.17)} \qquad \limsup_{t\to\infty} \sup_{x \geq -\zeta} \left| u(x,t) - \sum_{p=1}^{N} \left(-2\kappa_p^2 \operatorname{sech}^2[\kappa_p(x - x_p^+ - 4\kappa_p^2 t)]\right)\right| = 0$$

with x_p^+ *as in (5.21b).*

Furthermore, it follows from (7.15) and (3.73) that the nonsoliton part of the solution satisfies, in the coordinate region $t \geq t_{cd}$, $x \geq -\zeta$, $\zeta = \mu + \nu T$, $T = (3t)^{1/3}$ the x and t dependent bound

$$\text{(7.18a)} \qquad |u_c(x,t)| \leq \tilde{a} T^{-1} \exp\left[-\frac{2}{3}\theta\left(\frac{x+\zeta}{T}\right)^{3/2}\right] + \tilde{b} T^{-2}\left(1 + \frac{x+\zeta}{T}\right)^{-3/2}$$

$$\text{(7.18b)} \qquad \tilde{a} = (\tilde{\omega}_0 + \tilde{\omega}_1)\gamma_\theta, \qquad \tilde{b} = (\tilde{\omega}_0 + \tilde{\omega}_1)\rho,$$

with $\theta \in (0,1)$, γ_θ, ρ the constants introduced at the end of section 3. The estimate (7.18) is to be compared with the estimate (4.19) obtained in the absence of solitons. Clearly, by contrast with (4.20), the bound (7.18) does not permit us to conclude that the L^1 $(-\zeta \leq x < +\infty)$-norm of $u_c(x,t)$ tends to zero as $t \to \infty$.

However, we obtain from (2.28), (3.65a), (7.6)

$$\text{(7.19)} \qquad \int_{-\zeta}^{\infty} u_c(x,t)dx = \beta_c(0^+;-\zeta,t) = O(t^{-1/3}) \quad \text{as } t \to \infty.$$

Hence, in view of (5.19a-20a)

$$\text{(7.20a)} \qquad \int_{-\zeta}^{\infty} u(x,t)dx = -4\sum_{p=1}^{N} \kappa_p + O(t^{-1/3}) \qquad \text{as } t \to \infty.$$

From the formula (see [18])

$$(7.21)\qquad \int_{-\infty}^{\infty} u(x,t)dx = -\frac{2}{\pi}\int_0^{\infty} \log(1-|b_r(k)|^2)dk - 4\sum_{p=1}^{N}\kappa_p$$

we find for the complementary integral

$$(7.20b)\qquad \int_{-\infty}^{-\zeta} u(x,t)dx = -\frac{2}{\pi}\int_0^{\infty} \log(1-|b_r(k)|^2)dk + O(t^{-1/3})$$

as $t \to \infty$.

In particular, in the case of a nonzero reflection coefficient, it follows from (2.12), (7.20) that there exists a t_0 such that for $t \geq t_0$

$$(7.22)\qquad \int_{-\infty}^{0} u(x,t)dx > 0 \text{ and } \int_0^{\infty} u(x,t)dx < 0.$$

Thus, in the light of (5.18), the reflectionless solutions of the KdV equation can be characterized as the only nontrivial solutions that do not assume positive values.

Let us mention, that the time independent quantity (7.21) is usually referred to (cf. [17]) as the total momentum associated with the solution $u(x,t)$ of (7.14). For nonzero b_r, (7.20) suggests that as time goes on there is a definite positive momentum associated with the dispersive wavetrain given by (7.20b), as well as a definite negative momentum associated with the pure N-soliton solution given by (7.20a). Incidentally, observe that (7.21) gives an immediate proof of the following result which is partly known from quantum mechanics [11]:

Let $u_0(x)$ be an arbitrary potential in the Schrödinger scattering problem, satisfying the conditions of subsection 2.1.

If $\int_{-\infty}^{\infty} u_0(x)dx < 0$ then u_0 has at least one bound state; if $u_0 \not\equiv 0$ and $\int_{-\infty}^{\infty} u_0(x)dx = 0$, then u_0 has at least one bound state and a nonzero reflection coefficient; if $\int_{-\infty}^{\infty} u_0(x)dx > 0$ then u_0 has a nonzero reflection coefficient.

We continue our list of consequences of theorem 7.1 by considering L^2-estimates. From (7.18) we find

$$(7.23)\qquad \int_{-\zeta}^{\infty} u_c^2(x,t)dx = O(t^{-1/3}) \quad \text{as } t \to \infty.$$

Since by (5.20a) and (7.16)

$$(7.24) \qquad \int_{-\zeta}^{\infty} |u_c(x,t)u_d(x,t)|dx \leq 4\tilde{\omega}_2 T^{-1} \sum_{p=1}^{N} \kappa_p,$$

we conclude from (5.19b-20b) that

$$(7.25a) \qquad \int_{-\zeta}^{\infty} u^2(x,t)dx = \frac{16}{3}\sum_{p=1}^{N} \kappa_p^3 + O(t^{-1/3}) \quad \text{as } t \to \infty.$$

Using the formula (see [18])

$$(7.26) \qquad \int_{-\infty}^{\infty} u^2(x,t)dx = -\frac{8}{\pi}\int_{0}^{\infty} k^2 \log(1 - |b_r(k)|^2)dk + \frac{16}{3}\sum_{p=1}^{N} \kappa_p^3,$$

we obtain as a counterpart to (7.25a)

$$(7.25b) \qquad \int_{-\infty}^{-\zeta} u^2(x,t)dx = -\frac{8}{\pi}\int_{0}^{\infty} k^2 \log(1 - |b_r(k)|^2)dk + O(t^{-1/3}) \quad \text{as } t \to \infty.$$

In the literature [10] the time independent quantity (7.26), sometimes [17] with a factor ½ in front of it, is referred to as the energy associated with the solution u(x,t) of (7.14). For nonzero b_r, (7.25) suggests that as $t \to \infty$ the dispersive wavetrain moving to the left, though it may decay asymptotically to zero amplitude, still carries a finite amount of energy given by (7.25b), while on the other side of the line the N-soliton solution, falling apart into N solitons moving to the right, carries the energy given by (7.25a). It is interesting to compare the $O(t^{-1/3})$ term in (7.25b), due to the interaction between the dispersive wavetrain and the N solitons, with the $O(t^{-1})$ term in (4.22b) caused by the self-interaction of the dispersive wavetrain, in the absence of solitons.

Finally, let us remark that (7.16) improves Tanaka's result ([16], Theorem 1.1), which can be reformulated as

$$(7.27) \qquad \lim_{t\to\infty} \sup_{x \geq vt} |u_c(x,t)| = 0 \text{ for } v > 0 \text{ arbitrarily fixed.}$$

Moreover, it is not difficult to derive more precise versions of (7.27).

As a first example, let us make the additional assumption

(7.28) There is an integer $\tilde{n} \geq 2$ such that $b_r \in C^{\tilde{n}}(\mathbb{R})$ and all derivatives $b_r^{(j)}(k)$, $j = 0,1,\ldots,\tilde{n}$ satisfy

$$b_r^{(j)}(k) = \mathcal{O}(|k|^{-3}) \qquad k \to \pm\infty.$$

Then, it follows from a slight modification of Chapter 1, Appendix B, that, given the positive constant v, there is a constant μ_0 such that

$$|\Omega_c(\xi;t)| + \left|\frac{\partial}{\partial\xi}\Omega_c(\xi;t)\right| \leq \mu_0\xi^{-\tilde{n}} \text{ for } \xi \geq vt > 0. \tag{7.29}$$

Hence, by (7.15b) we can choose t_{cd} such that

$$|u_c(x,t)| \leq \tilde{\mu}_0 x^{-\tilde{n}}, \qquad t \geq t_{cd}, \qquad x \geq vt, \tag{7.30}$$

where $\tilde{\mu}_0$ is some constant. Thus we arrive at

$$\sup_{x \geq vt} |u_c(x,t)| = \mathcal{O}(t^{-\tilde{n}}) \qquad \text{as } t \to \infty. \tag{7.31}$$

Note that, at the same time, (7.31) improves some of the results obtained in Chapter 1, since it is easy to show that (7.29-30-31) with slightly different constants μ_0, $\tilde{\mu}_0$, t_{cd} are still valid if the condition (7.28) is relaxed to that stated in Chapter 1, Appendix B, with $n = \tilde{n}$.

Incidentally, we can apply the above results to estimate the decay rate of the solution u(x,t) of (7.14) as $x \to +\infty$ for fixed $t \geq t_{cd}$. Since, in view of (7.5-16c), $u_d(x,t)$ decays exponentially as $x \to +\infty$, we find from (7.15a) and (7.30) that for $t \geq t_{cd}$

$$u(x,t) = \mathcal{O}(x^{-\tilde{n}}) \qquad \text{as } x \to +\infty. \tag{7.32}$$

If u_0 satisfies (1.3), then we obtain from (2.22) that (7.28) holds with $\tilde{n} = [\![M]\!] + 2 - (\gamma/2)$. Hence, for $t \geq t_{cd}$

$$u(x,t) = \mathcal{O}(x^{(\gamma/2)-[\![M]\!]-2}) \qquad \text{as } x \to +\infty. \tag{7.33}$$

Herewith, for $t \geq t_{cd}$, the estimate

$$u(x,t) = \mathcal{O}(x^{(\gamma/2)-[\![M]\!]}) \qquad \text{as } x \to +\infty, \tag{7.34}$$

which was obtained in [4] for any fixed $t > 0$, is improved.

As a second example, let us suppose that u_0 satisfies (1.3) and furthermore, that there exists an $\varepsilon_0 > 0$ such that $u_0(x) = \mathcal{O}(\exp(-2\varepsilon_0 x))$ as $x \to +\infty$.

Since u_0 satisfies (1.3), we can apply theorem 7.1 to find a constant t_{cd} such that (7.15) holds for $t \geq t_{cd}$, $x \geq 0$. Now, fix $v > 0$. Then it follows from combining the last remark of subsection 2.1 with Chapter 1, Appendix A, that, given ε_1 with $0 < \varepsilon_1 < \min(\varepsilon_0, \kappa_N)$, there exists a constant γ_1 such that

$$(7.35) \qquad |\Omega_c(\xi;t)| + \left|\frac{\partial}{\partial\xi}\Omega_c(\xi;t)\right| \leq \gamma_1 \exp(-2\varepsilon_1\xi + 8\varepsilon_1^3 t), \; t \geq t_{cd}, \; \xi \geq vt.$$

Hence, by (7.15b)

$$(7.36) \qquad |u_c(x,t)| \leq \tilde{\gamma}_1 \exp(-2\varepsilon_1 x + 8\varepsilon_1^3 t), \quad t \geq t_{cd}, \quad x \geq vt$$

where $\tilde{\gamma}_1$ is some constant.

Firstly, (7.36) gives us the decay rate of the solution $u(x,t)$ of (7.14) as $x \to +\infty$ for fixed $t \geq t_{cd}$. Since, by (2.6) and (5.7-16c), $u_d(x,t) = \mathcal{O}(\exp(-2\kappa_N x))$ as $x \to +\infty$, we conclude from (7.15a) and (7.36) that for $t \geq t_{cd}$

$$(7.37) \qquad u(x,t) = \mathcal{O}(\exp(-2\varepsilon_1 x)) \quad \text{as } x \to +\infty \text{ for any } \varepsilon_1 \text{ with } 0 < \varepsilon_1 < \min(\varepsilon_0, \kappa_N).$$

Secondly, choosing ε_1 such that $0 < \varepsilon_1 < \min(\varepsilon_0, \kappa_N, \frac{1}{2}\sqrt{v})$, we obtain from (7.36)

$$(7.38) \qquad \sup_{x \geq vt} |u_c(x,t)| = \mathcal{O}(\exp(-\alpha_1 t)) \quad \text{as } t \to \infty$$

with $\alpha_1 = 2\varepsilon_1(v - 4\varepsilon_1^2) > 0$.

And so we have obtained another more precise version of (7.27).

References

[1] M.J. Ablowitz and H. Segur, Asymptotic solutions of the Korteweg-de Vries equation, Stud. Appl. Math. 57 (1977), 13-44.

[2] M.J. Ablowitz and H. Segur, Solitons and the Inverse Scattering Transform, Philadelphia, SIAM, 1981.

[3] M. Abramowitz and I.A. Stegun, Handbook of Mathematical Functions, National Bureau of Standards Applied Mathematics Series, No. 55, U.S. Department of Commerce, 1964.

[4] A. Cohen, Existence and regularity for solutions of the Korteweg-de Vries equation, Arch. for Rat. Mech. and Anal. 71 (1979), 143-175.

[5] P.J. Davis, Interpolation and Approximation, Dover, New York, 1963.

[6] P. Deift and E. Trubowitz, Inverse scattering on the line, Comm. Pure Appl. Math. 32 (1979), 121-251.

[7] W. Eckhaus and A. van Harten, The Inverse Scattering Transformation and the Theory of Solitons, North-Holland Mathematics Studies 50, 1981.

[3] W. Eckhaus and P. Schuur, The emergence of solitons of the Korteweg-de Vries equation from arbitrary initial conditions, Math. Meth. in the Appl. Sci. 5 (1983), 97-116.

[9] C.S. Gardner, J.M. Greene, M.D. Kruskal and R.M. Miura, Method for solving the Korteweg-de Vries equation, Phys. Rev. Lett. 19 (1967), 1095-1097.

[10] C.S. Gardner, J.M. Greene, M.D. Kruskal and R.M. Miura, Korteweg-de Vries equation and generalizations VI, Comm. Pure Appl. Math. 27 (1974), 97-133.

[11] L. Landau and E. Lifschitz, Quantum Mechanics, Nonrelativistic Theory, Pergamon Press, New York, 1958.

[12] P.D. Lax, Integrals of nonlinear equations of evolution and solitary waves, Comm. Pure Appl. Math. 21 (1968), 467-490.

[13] J.W. Miles, The asymptotic solution of the Korteweg-de Vries equation in the absence of solitons, Stud. Appl. Math. 60 (1979), 59-72.

[14] F.W. Olver, Asymptotics and Special Functions, Academic Press, New York, 1974.

[15] S. Tanaka, On the N-tuple wave solutions of the Korteweg-de Vries equation, Publ. R.I.M.S. Kyoto Univ. 8 (1972), 419-427.

[16] S. Tanaka, Korteweg-de Vries equation; asymptotic behavior of solutions, Publ. R.I.M.S. Kyoto Univ. 10 (1975), 367-379.

[17] N.J. Zabusky, Solitons and bound states of the time-independent Schrödinger equation, Phys. Rev. 168 (1968), 124-128.

[18] V.E. Zakharov and L.D. Faddeev, Korteweg-de Vries equation, a completely integrable Hamiltonian system, Funct. Anal. Appl. 5 (1971), 280-287.

CHAPTER THREE

MULTISOLITON PHASE SHIFTS FOR THE KORTEWEG-DE VRIES EQUATION IN THE CASE OF A NONZERO REFLECTION COEFFICIENT

We study multisoliton solutions of the Korteweg-de Vries equation in the case of a nonzero reflection coefficient. An explicit phase shift formula is derived that clearly displays the nature of the interaction of each soliton with the other ones and with the dispersive wavetrain. In particular, this formula shows that each soliton experiences in addition to the ordinary N-soliton phase shift an extra phase shift to the left caused by the collision with the dispersive wavetrain.

1. Introduction.

We consider the Korteweg-de Vries (KdV) equation $u_t - 6uu_x + u_{xxx} = 0$ with arbitrary real initial conditions $u(x,0) = u_0(x)$, which are sufficiently smooth and decay sufficiently rapidly for $|x| \to \infty$ for the whole of the inverse scattering method to work and to guarantee certain regularity and decay properties of the scattering data, to be stated further on. The long-time behaviour of the solution $u(x,t)$ of the KdV problem has been discussed by numerous authors. The general picture is, that as $t \to +\infty$ the solution decomposes into N solitons moving to the right and a dispersive wavetrain moving to the left. As $t \to -\infty$ the arrangement is reversed. The emergence of the N solitons as $t \to +\infty$ for rather arbitrary classes of initial conditions was demonstrated rigorously in [8], Chapter 1 in this volume (see also the discussion in [7]). Earlier - but less detailed and not widely known - results in that direction were given in [11]. Further extensions of the asymptotic analysis and improvements of results were recently presented in [10], Chapter 2 of the present volume. In the literature many attempts were made to calculate the phase shifts of the solitons as they interact both with the other solitons and with the dispersive wavetrain. Many incorrect results were given (cf. [11] and [12]), until finally the question was settled by Ablowitz and Kodama [1], who presented a correct phase shift formula.

In this chapter we rederive this phase shift formula, starting from our asymptotic analysis of the solution given in Chapter 2. We next show how a simple substitution produces a more transparent formula that clearly displays the nature of the interaction of each soliton with the other ones and with the dispersive wavetrain. From our phase shift formula it is evident, that each soliton experiences, in addition to the ordinary N-soliton phase shift, an extra phase shift to the left, the so-called continuous phase shift, caused by the collision with the dispersive wavetrain. Thus, the presence of reflection causes a delay in the soliton motion. Furthermore, our formula shows that the total phase shift is completely determined by the bound states and the right reflection coefficient. Hence, there is no dependence on the right normalization coefficients.
From the original formula the above facts are hard to see.

The composition of this chapter is as follows. In section 2 we briefly discuss the left and right scattering data associated with $u_0(x)$ and show how the left scattering data can be expressed in terms of the right scattering data in a convenient way. In section 3 we recall a result known from Chapter 2, concerning the asymptotic behaviour of $u(x,t)$ as $t \to +\infty$. By a symmetry argument we derive from this result the asymptotic behaviour of $u(x,t)$ as $t \to -\infty$. Next, in section 4, the two asymptotic results are combined to give the Ablowitz-Kodama phase shift formula. The representation of the left normalization coefficients in terms of the right scattering data, which was obtained in section 2, then enables us to write the phase shift formula in a more transparent form. Finally, as an exercise, we calculate in section 5 the continuous phase shifts arising from a sech^2 initial function.

2. Scattering data and their properties.

For Im $k \geq 0$ we introduce the Jost functions $\psi_r(x,k)$ and $\psi_\ell(x,k)$, two special solutions of the Schrödinger equation

$$\psi_{xx} + (k^2 - u_0(x))\psi = 0, \qquad -\infty < x < +\infty \tag{2.1}$$

determined by

$$\psi_r(x,k) = e^{-ikx}R(x,k), \qquad \lim_{x\to-\infty} R(x,k) = 1, \qquad \lim_{x\to-\infty} R_x(x,k) = 0 \tag{2.2a}$$

$$\psi_\ell(x,k) = e^{ikx}L(x,k), \qquad \lim_{x\to+\infty} L(x,k) = 1, \qquad \lim_{x\to+\infty} L_x(x,k) = 0. \tag{2.2b}$$

We set

$$r_-(k) = 1 - (2ik)^{-1} \int_{-\infty}^{\infty} u_0(y)R(y,k)dy \qquad k \in \overline{\mathbb{C}}_+\backslash\{0\} \tag{2.3a}$$

$$r_+(k) = (2ik)^{-1} \int_{-\infty}^{\infty} e^{-2iky}u_0(y)R(y,k)dy \qquad k \in \mathbb{R}\backslash\{0\} \tag{2.3b}$$

$$\ell_+(k) = 1 - (2ik)^{-1} \int_{-\infty}^{\infty} u_0(y)L(y,k)dy \qquad k \in \overline{\mathbb{C}}_+\backslash\{0\} \tag{2.3c}$$

$$\ell_-(k) = (2ik)^{-1} \int_{-\infty}^{\infty} e^{2iky}u_0(y)L(y,k)dy \qquad k \in \mathbb{R}\backslash\{0\}. \tag{2.3d}$$

Note that $r_-(k) = \ell_+(k)$, whereas $r_+(k) = -\ell_-(-k)$. It is well known [7], that $r_-(k)$ is analytic on $\mathbb{C}_+$ with at most finitely many zeros, all simple and on the imaginary axis. Let us denote them by $i\kappa_m$, $m = 1,2,\dots,N$ and order

$$\kappa_1 > \kappa_2 > \dots > \kappa_N > 0. \tag{2.4}$$

Bearing in mind that $\psi_\ell(x,i\kappa_m)$ and $\psi_r(x,i\kappa_m)$ are both real-valued and square integrable, we introduce

(2.5a) $$c_m^r = \left[\int_{-\infty}^{\infty} \psi_\ell^2(x,i\kappa_m)dx\right]^{-\frac{1}{2}},$$ the right normalization coefficients,

(2.5b) $$c_m^\ell = \left[\int_{-\infty}^{\infty} \psi_r^2(x,i\kappa_m)dx\right]^{-\frac{1}{2}},$$ the left normalization coefficients.

Furthermore, we introduce the following quantities for $k \in \mathbb{R}\backslash\{0\}$

(2.6a) $a_r = r_-^{-1}$, the right transmission coefficient

(2.6b) $a_\ell = \ell_+^{-1}$, the left transmission coefficient,

(2.6c) $b_r = r_+ r_-^{-1}$, the right reflection coefficient

(2.6d) $b_\ell = \ell_- \ell_+^{-1}$, the left reflection coefficient.

Assuming that $u_0(x)$ decays sufficiently rapidly (see [7]) we can extend a_r, a_ℓ, b_r, b_ℓ in a natural way to continuous functions on all of $\mathbb{R}$. We shall call the aggregate of quantities $\{a_r(k), b_r(k), \kappa_m, c_m^r\}$ the right scattering data of the potential u_0. Similarly we refer to $\{a_\ell(k), b_\ell(k), \kappa_m, c_m^\ell\}$ as the left scattering data associated with u_0. In a different, but equivalent way the right scattering data were already introduced in Chapter 2, section 2 (see also [7], Ch. 4).

We claim that a_ℓ, b_ℓ and c_m^ℓ can be expressed in terms of the right scattering data in the following way

(2.7a) $$a_\ell(k) = a_r(k), \qquad b_\ell(k) = -\frac{a_r(k)}{a_r(-k)}\, b_r(-k),$$

(2.7b) $$c_m^\ell = [c_m^r]^{-1} 2\kappa_m \left\{\exp\left\{\frac{\kappa_m}{\pi} \int_0^\infty \frac{\log(1-|b_r(k)|^2)}{k^2+\kappa_m^2}\, dk\right\}\right\} \prod_{\substack{p=1\\p\neq m}}^{N} \left|\frac{\kappa_m+\kappa_p}{\kappa_m-\kappa_p}\right| .$$

Indeed, the relations (2.7a) are obvious. To derive (2.7b) we combine certain familiar facts from [6], [7]. Firstly, from [7], p. 110 we know

(2.8) $$\psi_r(x,i\kappa_m) = \alpha_m \psi_\ell(x,i\kappa_m), \quad \text{with } \alpha_m \in \mathbb{R}\backslash\{0\}.$$

Hence, by (2.5)

(2.9) $$c_m^r = |\alpha_m| c_m^\ell .$$

Next, by [7], (4.3.18) one has

(2.10) $$\left.\frac{dr_-}{dk}\right|_{k=i\kappa_m} = (i\alpha_m)^{-1} \int_{-\infty}^{\infty} \psi_r^2(x,i\kappa_m)dx = (i\alpha_m)^{-1}[c_m^\ell]^{-2}.$$

Eliminating α_m from (2.9) and (2.10) we find

(2.11) $$c_m^r c_m^\ell \left| \left.\frac{dr_-}{dk}\right|_{k=i\kappa_m} \right| = 1.$$

Lastly, from [6], p. 154 we obtain the representation

(2.12) $$r_-(k) = \left\{\exp\left\{\frac{1}{2\pi i} \int_{-\infty}^{\infty} \frac{\log(1-|b_r(\omega)|^2)}{k-\omega} d\omega\right\}\right\} \prod_{p=1}^{N} \frac{k-i\kappa_p}{k+i\kappa_p} , \quad \text{Im } k > 0.$$

Consequently

(2.13) $$\left| \left.\frac{dr_-}{dk}\right|_{k=i\kappa_m} \right| = \frac{1}{2\kappa_m}\left\{\exp\left\{-\frac{\kappa_m}{\pi} \int_0^{\infty} \frac{\log(1-|b_r(k)|^2)}{k^2+\kappa_m^2} dk\right\}\right\} \prod_{\substack{p=1 \\ p\neq m}}^{N} \left|\frac{\kappa_m-\kappa_p}{\kappa_m+\kappa_p}\right| ,$$

where we have used that $b_r^*(k) = b_r(-k)$.
Combining (2.11) and (2.13) we arrive at the desired formula (2.7b).

3. Forward and backward asymptotics.

Once the right scattering data of $u_0(x)$ are known, the solution $u(x,t)$ of the forward KdV problem

(3.1) $$\begin{cases} u_t - 6uu_x + u_{xxx} = 0, & t > 0 \\ u(x,0) = u_0(x) \end{cases}$$

can in principle be computed by the inverse scattering method [7]. Concerning the asymptotic behaviour of the solution we have obtained the following result in Chapter 2, section 7.

Lemma 3.1. *Assume that*

(3.2) $\quad b_r(k)$ *is of class* $C^2(\mathbb{R})$ *and the derivatives* $b_r^{(j)}(k)$, $j = 0,1,2$ *satisfy*

$$b_r^{(j)}(k) = O(|k|^{-1}), \qquad k \to \pm\infty.$$

Then one has

$$\lim_{t\to\infty} \sup_{x \geq -t^{1/3}} \left| u(x,t) - \sum_{m=1}^{N} \left(-2\kappa_m^2 \operatorname{sech}^2[\kappa_m(x - x_m^+ - 4\kappa_m^2 t)]\right) \right| = 0, \tag{3.3}$$

where

$$x_m^+ = \frac{1}{2\kappa_m} \log\left\{\frac{[c_m^r]^2}{2\kappa_m} \prod_{p=1}^{m-1} \left(\frac{\kappa_p - \kappa_m}{\kappa_p + \kappa_m}\right)^2\right\}. \tag{3.4}$$

Let us now consider the backward KdV problem, starting from the same initial function $u_0(x)$, i.e.

$$\begin{cases} u_t - 6uu_x + u_{xxx} = 0, & t < 0 \\ u(x,0) = u_0(x). \end{cases} \tag{3.5}$$

Clearly, if $u(x,t)$ satisfies (3.5), then $w(x,t) = u(-x,-t)$ satisfies

$$\begin{cases} w_t - 6ww_x + w_{xxx} = 0, & t > 0 \\ w(x,0) = u_0(-x), \end{cases} \tag{3.6}$$

so that $w(x,t)$ satisfies the forward KdV problem with initial function $u_0(-x)$. To solve (3.5) it is therefore sufficient to determine the right scattering data associated with $u_0(-x)$ and apply the inverse scattering method to (3.6). However, it is readily verified that the right scattering data associated with $u_0(-x)$ are equal to the left scattering data associated with $u_0(x)$, which were studied in the previous section. Thus, to find the asymptotic behaviour of the solution $u(x,t)$ of (3.5) for $t \to -\infty$ we merely apply lemma 3.1 to problem (3.6) and perform the transcription $u(x,t) = w(-x,-t)$. This yields

Lemma 3.2. *Assume that*

(3.7) $b_\ell(k)$ *is of class* $C^2(\mathbb{R})$ *and the derivatives* $b_\ell^{(j)}(k)$, $j = 0,1,2$ *satisfy*

$$b_\ell^{(j)}(k) = O(|k|^{-1}), \qquad k \to \pm\infty.$$

Then one has

$$\lim_{t\to-\infty} \sup_{x\le|t|^{1/3}} \left|u(x,t) - \sum_{m=1}^{N} (-2\kappa_m^2 \operatorname{sech}^2[\kappa_m(x-x_m^- - 4\kappa_m^2 t)])\right| = 0, \tag{3.8}$$

where

$$x_m^- = -\frac{1}{2\kappa_m} \log\left\{\frac{[c_m^\ell]^2}{2\kappa_m} \prod_{p=1}^{m-1} \left(\frac{\kappa_p-\kappa_m}{\kappa_p+\kappa_m}\right)^2\right\}. \tag{3.9}$$

4. An explicit phase shift formula.

Let us assume that b_r and b_ℓ satisfy the conditions (3.2) and (3.7). Then the convergence results (3.3) and (3.8) display clearly how the solution $u(x,t)$ of the KdV equation evolving from $u(x,0) = u_0(x)$ splits up into N solitons as $t \to \pm\infty$.

In particular, we find for the m-th soliton the following phase shift

$$S_m = x_m^+ - x_m^- = \frac{1}{2\kappa_m} \log\left\{\left(\frac{c_m^r c_m^\ell}{2\kappa_m}\right)^2 \prod_{p=1}^{m-1} \left(\frac{\kappa_m-\kappa_p}{\kappa_m+\kappa_p}\right)^4\right\}. \tag{4.1}$$

This formula was first derived by Ablowitz and Segur [2] for the N = 1 case and by Ablowitz and Kodama [1] for the N > 1 case (see also the discussion in [3]).

It is a remarkable fact that the formulae (3.9) and (4.1) become both more transparent and more meaningful if one inserts the representation (2.7b). Summarizing, this leads to

$$x_m^+ = \frac{1}{2\kappa_m} \log\left(\frac{[c_m^r]^2}{2\kappa_m}\right) + \frac{1}{\kappa_m} \sum_{p=1}^{m-1} \log\left(\frac{\kappa_p-\kappa_m}{\kappa_p+\kappa_m}\right) \tag{4.2a}$$

$$x_m^- = \frac{1}{2\kappa_m} \log\left(\frac{[c_m^r]^2}{2\kappa_m}\right) + \frac{1}{\kappa_m} \sum_{p=m+1}^{N} \log\left(\frac{\kappa_m-\kappa_p}{\kappa_m+\kappa_p}\right) - \frac{1}{\pi} \int_0^\infty \frac{\log(1-|b_r(k)|^2)}{k^2+\kappa_m^2}\, dk \tag{4.2b}$$

(4.3a) $$S_m = S_m^d + S_m^c$$

(4.3b) $$S_m^d = \frac{1}{\kappa_m} \sum_{p=1}^{m-1} \log\left(\frac{\kappa_p - \kappa_m}{\kappa_p + \kappa_m}\right) - \frac{1}{\kappa_m} \sum_{p=m+1}^{N} \log\left(\frac{\kappa_m - \kappa_p}{\kappa_m + \kappa_p}\right)$$

(4.3c) $$S_m^c = \frac{1}{\pi} \int_0^\infty \frac{\log(1 - |b_r(k)|^2)}{k^2 + \kappa_m^2} dk.$$

In S_m^d we recognize the pure N-soliton phase shift (caused by pairwise interaction of the m-th soliton with the other ones). The quantity S_m^c (which is *negative* for nonzero b_r) can be seen as the shift caused by the interaction of the m-th soliton with the dispersive wavetrain. Note that the phase shift S_m is completely determined by the bound states κ_p and the right reflection coefficient b_r and is thus independent of the right normalization coefficients, a fact not in the least suggested by the original formula (4.1). For nonzero b_r we obviously have

(4.4) $$0 > S_1^c > S_2^c > \ldots > S_N^c.$$

Thus, the collision with the dispersive wavetrain causes a delay in the motion of the solitons and the effect is most heavily felt by the smallest one, corresponding to κ_N.
Using the formula (see [3])

(4.5) $$\int_{-\infty}^{\infty} u_0(x)dx = -\frac{2}{\pi} \int_0^\infty \log(1 - |b_r(k)|^2)dk - 4 \sum_{p=1}^{N} \kappa_p$$

we obtain for the continuous phase shift S_m^c the following estimate in terms of the initial function $u_0(x)$ and the bound states κ_p:

(4.6) $$|S_m^c| \leq \frac{1}{2\kappa_m^2} \left(\int_{-\infty}^{\infty} u_0(x)dx + 4 \sum_{p=1}^{N} \kappa_p \right).$$

In estimating the size of S_m^c one has to distinguish two cases, the "generic case" and the "exceptional case" (see [5], [6], as well as Chapter 2, subsection 2.1). In the generic case, the Jost functions $\psi_r(x,0)$ and $\psi_\ell(x,0)$ are linearly independent, whereas in the exceptional case they are not. In the exceptional case one has

(4.7) $$B = \sup_{k \in \mathbb{R}} |b_r(k)| < 1,$$

whence

(4.8) $$|S_m^c| \leq -\frac{1}{2\kappa_m}\log(1 - B^2).$$

In the generic case there is an $\alpha \neq 0$ with

(4.9) $$b_r(k) = -1 + \alpha k + o(|k|) \text{ as } k \to 0,$$

so that in the integral defining S_m^c the contribution of $k = 0$ becomes important. In particular, fixing $|b_r|$, we find for $\kappa_m \downarrow 0$

(4.10a) $$S_m^c \sim \frac{1}{2\kappa_m}\log(1 - |b_r(0)|^2) \quad \text{in the exceptional case}$$

(4.10b) $$S_m^c \sim \frac{1}{\kappa_m}\log \kappa_m \quad \text{in the generic case.}$$

Clearly, in general the sizes of S_m^c and S_m^d are incomparable.
On the other hand one can easily construct examples in which one of the two dominates. For instance, consider a generic case with two bound states κ_1 and $\kappa_2 = \frac{1}{2}\kappa_1$. Then, for fixed $|b_r|$, the discrete phase shifts S_m^d dominate for $\kappa_1 \to +\infty$, whereas the continuous phase shifts S_m^c dominate for $\kappa_1 \downarrow 0$; in the $\kappa_1 \downarrow 0$ case the familiar picture of a KdV soliton overtaking a smaller one, where the smaller one is shifted to the left and the larger one to the right, changes, since now both are shifted to the left.

5. An example: the continuous phase shifts arising from a sech² initial function.

To illustrate the previous discussion let us compute the continuous phase shifts arising from the initial function

(5.1) $$u_0(x) = -\lambda(\lambda+1)\text{sech}^2 x, \qquad \lambda > 0.$$

From [9] we find

(5.2a) $$a_r(k) = \frac{\Gamma(a)\Gamma(b)}{\Gamma(c)\Gamma(a+b-c)}, \qquad b_r(k) = \frac{\Gamma(c-a-b)\Gamma(a)\Gamma(b)}{\Gamma(c-a)\Gamma(c-b)\Gamma(a+b-c)}, \text{ with}$$

(5.2b) $$a = 1 + \lambda - ik, \quad b = -\lambda - ik, \quad c = 1 - ik,$$

where Γ denotes the gamma function ([4], p. 253). Clearly, a_r is analytic on $C_+\setminus\{\kappa_1,\kappa_2,\ldots,\kappa_N\}$ with simple poles at the bound states $\kappa_1,\kappa_2,\ldots,\kappa_N$. Here $N \geqq 1$ is the unique integer such that $N-1 < \lambda \leqq N$ and the κ_p are given by

$$(5.3) \qquad \kappa_p = 1 + \lambda - p, \quad p = 1,2,\ldots,N.$$

Note, that $u_0(x)$ is reflectionless (i.e. $b_r \equiv 0$) if and only if $\lambda = 1,2,\ldots$, in which case $N = \lambda$. For the other values of λ we find that $b_r(0) = -1$ so that we are in the generic case.
To compute the continuous phase shifts S_m^c we notice that by (5.2)

$$(5.4) \qquad r_-(i\nu) = \frac{\Gamma(\nu)\Gamma(1+\nu)}{\Gamma(1+\lambda+\nu)\Gamma(\nu-\lambda)}, \quad \nu > 0.$$

On the other hand, by (2.12)

$$(5.5) \qquad r_-(i\nu) = \left\{\exp\left\{-\frac{\nu}{\pi}\int_0^\infty \frac{\log(1-|b_r(k)|^2)}{k^2+\nu^2}\,dk\right\}\right\} \prod_{p=1}^{N} \frac{\nu-\kappa_p}{\nu+\kappa_p}, \quad \nu > 0.$$

Equating both expressions we obtain, after repeated use of the recurrence formula $\Gamma(z+1) = z\Gamma(z)$, the following identity

$$(5.6) \qquad \frac{1}{\pi}\int_0^\infty \frac{\log(1-|b_r(k)|^2)}{k^2+\nu^2}\,dk = -\frac{1}{\nu}\log\left\{\frac{\Gamma(\nu)\Gamma(1+\nu)}{\Gamma(1+\nu+\lambda-N)\Gamma(\nu-\lambda+N)}\right\}$$

$$= -\frac{1}{\nu}\log\left\{\frac{B(\nu,1+\nu)}{B(1+\nu+\lambda-N,\nu-\lambda+N)}\right\}$$

where B refers to the beta function ([4], p. 258).
Finally, combining (5.3) and (5.6), we find that the continuous phase shifts S_m^c are given by

$$(5.7) \qquad S_m^c = -\frac{1}{1+\lambda-m}\log\left\{\frac{B(2-m+\lambda,1-m+\lambda)}{B(2-m+2\lambda-N,1-m+N)}\right\}, \quad m = 1,2,\ldots,N.$$

However, to get an idea of the magnitude of S_m^c it is much simpler to employ the estimate (4.6) which gives us immediately

$$(5.8) \qquad |S_m^c| \leq \frac{(N-\lambda)(1+\lambda-N)}{(1+\lambda-m)^2}.$$

References

[1] M.J. Ablowitz, Y. Kodama: Note on asymptotic solutions of the Korteweg-de Vries equation with solitons. Stud. Appl. Math. 66 (1982) No. 2, 159-170.

[2] M.J. Ablowitz, H. Segur: Asymptotic Solutions of the Korteweg-de Vries Equation. Stud. Appl. Math. 57 (1977), 13-44.

[3] M.J. Ablowitz, H. Segur: Solitons and the Inverse Scattering Transform, Philadelphia, SIAM, 1981.

[4] M. Abramowitz, I.A. Stegun: Handbook of mathematical functions. National Bureau of Standards Applied Mathematics Series, No. 55. U.S. Department of Commerce, 1964.

[5] A. Cohen: Existence and Regularity for Solutions of the Korteweg-de Vries equation. Arch. for Rat. Mech. and Anal. 71 (1979), 143-175.

[6] P. Deift, E. Trubowitz: Inverse scattering on the line. Comm. Pure Appl. Math. 32 (1979, 121-251.

[7] W. Eckhaus, A. van Harten: The inverse scattering transformation and the theory of solitons. North-Holland Mathematics Studies 50, 1981.

[8] W. Eckhaus, P. Schuur: The Emergence of Solitons of the Korteweg-de Vries Equation from Arbitrary Initial Conditions. Math. Meth. in the Appl. Sci. 5 (1983), 97-116.

[9] G.L. Lamb Jr.: Elements of soliton theory. Wiley-Interscience, 1980.

[10] P. Schuur: Asymptotic estimates of solutions of the Korteweg-de Vries equation on right half lines slowly moving to the left, preprint 330, Mathematical Institute Utrecht (1984).

[11] S. Tanaka: Korteweg-de Vries Equation; Asymptotic Behavior of Solutions. Publ. R.I.M.S. Kyoto Univ. 10 (1975), 367-379.

[12] V.E. Zakharov: Kinetic equation for solitons. Soviet Phys. JETP 33 (1971), 538-541.

CHAPTER FOUR

ON THE APPROXIMATION OF A REAL POTENTIAL IN THE ZAKHAROV-SHABAT SYSTEM BY ITS REFLECTIONLESS PART

In this chapter the inverse scattering algorithm associated with the Zakharov-Shabat system with real potential is simplified considerably. Exploiting this simplification we derive an estimate which clearly displays how well the potential is approximated by its reflectionless part.

1. Introduction.

The inverse scattering method associated with the Zakharov-Shabat system with real potential [14] can be used to solve a rich class of integrable nonlinear evolution equations, counting the modified Korteweg-de Vries equation and the sine-Gordon equation among its most distinguished members (cf. [2], [9], [13]). However, the only solutions of these equations that can be computed in explicit form are the so-called reflectionless solutions, i.e. solutions whose associated right reflection coefficient is zero. In a more general setting this situation leads automatically to the following question: Given an arbitrary real potential in the Zakharov-Shabat system, in which sense is it approximated by its reflectionless part?

In this chapter we shall give an answer to this question. To this end we first simplify the inverse scattering algorithm by showing how the Gel'fand-Levitan equation that appears in the literature can be simplified to a scalar integral equation containing only a single integral. The newly found Gel'fand-Levitan operator has, when considered in the complex Hilbert space $L^2(0,\infty)$, the remarkable structure of the identity plus an antisymmetric operator. Exploiting this structure we shall derive a pointwise estimate of the difference between the potential and its reflectionless part, which is remarkably simple in form and depends only on the bound states and the right reflection coefficient associated with the potential.

Let us emphasize that in applications of the inverse scattering method (cf. [3], [9]) the scattering data are usually known in explicit form. Therefore this estimate has immediate consequences in a practical case.

In Chapter 5 and Chapter 7 we shall use our estimate for an asymptotic analysis of the modified Korteweg-de Vries and the sine-Gordon equation respectively.

The chapter runs as follows.
In section 2 we review the direct scattering problem for the Zakharov-Shabat system with real potential. In section 3 the inverse problem is discussed and simplified. Next, in section 4 we state our main result, which, after the introduction of a convenient notation and the derivation of a useful lemma in section 5, is proven in section 6.

2. Construction and properties of the scattering data.

Let us briefly discuss the direct scattering problem for the Zakharov-Shabat system

(2.1) $$\begin{pmatrix}\psi_1\\ \psi_2\end{pmatrix}' = \begin{pmatrix}-i\zeta & q\\ -q & i\zeta\end{pmatrix}\begin{pmatrix}\psi_1\\ \psi_2\end{pmatrix}, \quad ' = \frac{d}{dx}, \quad -\infty < x < +\infty$$

where $q = q(x)$ is a real function and ζ a complex parameter. For details and proofs we refer to [1], [2], [6], [12]. Our notation is similar to that used in [6].

Following [6] we assume that the potential q satisfies the hypotheses:

(2.2a) $$q \in C^1(\mathbb{R})$$

(2.2b) $$\lim_{|x|\to\infty} q(x) = \lim_{|x|\to\infty} q'(x) = 0$$

(2.2c) $$\int_{-\infty}^{\infty} (|q(s)| + |q'(s)|)ds < +\infty.$$

In addition we shall need some conditions on the zeros of the Wronskian of the right and left Jost solutions, to be specified presently in (2.13).

For Im $\zeta \geq 0$ we define the (right and left) Jost solutions $\psi_r(x,\zeta)$ and $\psi_\ell(x,\zeta)$ as the special solutions of (2.1) uniquely determined by

(2.3a) $$\psi_r(x,\zeta) = e^{-i\zeta x}R(x,\zeta), \qquad \lim_{x\to-\infty} R(x,\zeta) = \begin{pmatrix}1\\0\end{pmatrix}$$

(2.3b) $$\psi_\ell(x,\zeta) = e^{i\zeta x}L(x,\zeta), \qquad \lim_{x\to+\infty} L(x,\zeta) = \begin{pmatrix}0\\1\end{pmatrix}.$$

The vector functions R and L are continuous in (x,ζ) on $\mathbb{R}\times\overline{\mathbb{C}}_+$ and analytic in ζ on $\mathbb{C}_+$ for each $x \in \mathbb{R}$. Furthermore their components satisfy

(2.4) $$\max\left[\sup_{\mathbb{R}\times\overline{\mathbb{C}}_+} |R_i(x,\zeta)|, \sup_{\mathbb{R}\times\overline{\mathbb{C}}_+} |L_i(x,\zeta)|\right] \leq \exp\left\{\int_{-\infty}^{\infty} |q(s)|ds\right\},$$

$$i = 1,2.$$

For Im $\zeta \leq 0$ we set

(2.5a) $$\tilde{\psi}_r(x,\zeta) \equiv \begin{pmatrix}-\psi_{r_2}(x,-\zeta)\\ \psi_{r_1}(x,-\zeta)\end{pmatrix} = \begin{pmatrix}-\psi^*_{r_2}(x,\zeta^*)\\ \psi^*_{r_1}(x,\zeta^*)\end{pmatrix}$$

$$(2.5b)\qquad \tilde{\psi}_\ell(x,\zeta) \equiv \begin{pmatrix} \psi_{\ell_2}(x,-\zeta) \\ -\psi_{\ell_1}(x,-\zeta) \end{pmatrix} = \begin{pmatrix} \psi^*_{\ell_2}(x,\zeta^*) \\ -\psi^*_{\ell_1}(x,\zeta^*) \end{pmatrix}.$$

It is readily verified that $\tilde{\psi}_r$ and $\tilde{\psi}_\ell$ are solutions of (2.1). Moreover, for $x,\ \zeta \in \mathbb{R}$ one has

$$(2.6a)\qquad W(\psi_r,\tilde{\psi}_r) = |R_1(x,\zeta)|^2 + |R_2(x,\zeta)|^2 = 1$$

$$(2.6b)\qquad W(\tilde{\psi}_\ell,\psi_\ell) = |L_1(x,\zeta)|^2 + |L_2(x,\zeta)|^2 = 1,$$

where $W(\psi,\phi) = \psi_1\phi_2 - \psi_2\phi_1$ denotes the Wronskian of ψ and ϕ. Hence, for ζ real, the pairs $\psi_r, \tilde{\psi}_r$ and ψ_ℓ, $\tilde{\psi}_\ell$ constitute fundamental systems of solutions of equation (2.1). In particular, we have for $x,\ \zeta \in \mathbb{R}$

$$(2.7a)\qquad \psi_r(x,\zeta) = r_+(\zeta)\psi_\ell(x,\zeta) + r_-(\zeta)\tilde{\psi}_\ell(x,\zeta)$$

$$(2.7b)\qquad r_+(\zeta) = W(\tilde{\psi}_\ell,\psi_r)$$

$$(2.7c)\qquad r_-(\zeta) = W(\psi_r,\psi_\ell).$$

The representation (2.7c) makes it possible to extend $r_-(\zeta)$ to a function analytic on $\operatorname{Im}\zeta > 0$ and continuous on $\operatorname{Im}\zeta \geq 0$. The following properties are easily demonstrated:

$$(2.8a)\qquad |r_+(\zeta)|^2 + |r_-(\zeta)|^2 = 1, \qquad \zeta \in \mathbb{R}$$

$$(2.8b)\qquad r_+^*(\zeta) = r_+(-\zeta), \qquad \zeta \in \mathbb{R}$$

$$(2.8c)\qquad r_-^*(\zeta) = r_-(-\zeta^*), \qquad \operatorname{Im}\zeta \geq 0.$$

Furthermore one can derive the integral representations

$$(2.9a)\qquad r_+(\zeta) = -\int_{-\infty}^{\infty} q(s)e^{-2i\zeta s}R_1(s,\zeta)ds, \qquad \zeta \in \mathbb{R}$$

$$(2.9b)\qquad r_-(\zeta) = 1 + \int_{-\infty}^{\infty} q(s)R_2(s,\zeta)ds, \qquad \operatorname{Im}\zeta \geq 0.$$

In combination with (2.6a) these yield

$$(2.10)\qquad \max\left[\sup_{\zeta\in\mathbb{R}} |r_+(\zeta)|,\ \sup_{\zeta\in\mathbb{R}} |1 - r_-(\zeta)|\right] \leq \int_{-\infty}^{\infty} |q(s)|ds.$$

At $\zeta = 0$, the scattering problem (2.1) can be solved in closed form. One has

(2.11) $$R_1(x,0) = \cos\left[\int_{-\infty}^{x} q(s)ds\right], \qquad R_2(x,0) = -\sin\left[\int_{-\infty}^{x} q(s)ds\right]$$

so that by (2.9)

(2.12) $$r_+(0) = -\sin\left[\int_{-\infty}^{\infty} q(s)ds\right], \qquad r_-(0) = \cos\left[\int_{-\infty}^{\infty} q(s)ds\right].$$

In terms of r_- we make our final assumptions:

(2.13a) $r_-(\zeta) \neq 0$ for $\zeta \in \mathbb{R}$

(2.13b) All zeros of r_- in $\mathbb{C}_+$ are simple.

Let us emphasize that, strictly spoken, condition (2.13b) is not necessary. In fact, a very elegant direct and inverse scattering formalism using only (2.2) and (2.13a) has been developed by Tanaka in [12]. Our motivation for requiring (2.13b) is that it simplifies the reasoning considerably.

Condition (2.13a), on the other hand, cannot be omitted. It poses an implicit restriction on the potential q and forms therefore a weak point in the Zakharov-Shabat scattering theory, as developed so far. An obvious consequence of (2.13a) is, in view of (2.12),

(2.14) $$\int_{-\infty}^{\infty} q(s)ds \neq (k + \tfrac{1}{2})\pi, \qquad k \in \mathbb{Z}.$$

Other explicit consequences for the potential q are still to be found. Note, however, that for small potentials no problems arise. Specifically, if

(2.15) $$\int_{-\infty}^{\infty} |q(s)|ds < 1$$

then (2.10) shows that (2.13a) is fulfilled.

Moreover, if

(2.16) $$\int_{-\infty}^{\infty} |q(s)|ds < 0.904$$

then (2.13a) and (2.13b) are trivially fulfilled since $r_-(\zeta) \neq 0$ for $\mathrm{Im}\,\zeta \geq 0$ (see [2] for a proof).

We now proceed with the construction of the scattering data associated with q(x). As a result of (2.13a) the function $r_-(\zeta)$ has at most finitely many zeros $\zeta_1,\zeta_2,\ldots,\zeta_N$, $\operatorname{Im}\zeta_j > 0$. By (2.8c-13b) they are all simple and distributed symmetrically with respect to the imaginary axis. Let P be the number of purely imaginary zeros and set $M = (N - P)/2$. We order the ζ_j in such a way that

$$\zeta_{\sigma(j)} = -\zeta_j^*, \quad j = 1,2,\ldots,N, \tag{2.17}$$

where σ denotes the permutation among integers between 1 and N defined by

$$\begin{aligned} \sigma(j) &= j + 1 && j \text{ odd } \leqq 2M \\ &= j - 1 && j \text{ even } \leqq 2M \\ &= j && j > 2M. \end{aligned} \tag{2.18}$$

It is a remarkable fact, that the ζ_j are precisely the eigenvalues of (2.1) in the upper half plane (the so-called bound states). The associated L^2-eigenspaces are one-dimensional and spanned by the exponentially decaying vector functions $\psi_\ell(x,\zeta_j)$, $j = 1,2,\ldots,N$. Note that by (2.7c) there are nonzero constants $\alpha(\zeta_j)$ such that

$$\psi_r(x,\zeta_j) = \alpha(\zeta_j)\psi_\ell(x,\zeta_j). \tag{2.19}$$

One can now derive the representation

$$\frac{dr_-}{d\zeta}(\zeta_j) = -2i\alpha(\zeta_j)\int_{-\infty}^{\infty}\psi_{\ell_1}(s,\zeta_j)\psi_{\ell_2}(s,\zeta_j)ds. \tag{2.20}$$

Bearing in mind that the integral on the right does not vanish because of (2.13b), we define the (right) normalization coefficients by

$$c_j^r = \tfrac{1}{2}i\left[\int_{-\infty}^{\infty}\psi_{\ell_1}(s,\zeta_j)\psi_{\ell_2}(s,\zeta_j)ds\right]^{-1}. \tag{2.21}$$

It is easily seen that they satisfy the same symmetry relation as the ζ_j, i.e.

$$c_{\sigma(j)}^r = -(c_j^r)^*. \tag{2.22}$$

Next, we introduce the following functions of $\zeta \in \mathbb{R}$

(2.23a) $\quad a_r(\zeta) = 1/r_-(\zeta)$, the (right) transmission coefficient

(2.23b) $b_r(\zeta) = r_+(\zeta)/r_-(\zeta)$, the (right) reflection coefficient.

By (2.8) one has for $\zeta \in \mathbb{R}$

(2.24a) $a_r^*(\zeta) = a_r(-\zeta)$, $b_r^*(\zeta) = b_r(-\zeta)$

(2.24b) $|a_r(\zeta)|^2 - |b_r(\zeta)|^2 = 1$.

In [6] it is shown that b_r is an element of $C \cap L^1 \cap L^2(\mathbb{R})$, which behaves as $o(|\zeta|^{-1})$ for $\zeta \to \pm\infty$. Furthermore, it is an obvious consequence of (2.12) that

(2.25) $$b_r(0) = -\tan\left[\int_{-\infty}^{\infty} q(s)ds\right],$$

which shows that in general the reflection coefficient may assume arbitrarily large values.

Clearly, by imposing stronger regularity and decay conditions on $q(x)$ in addition to (2.2-13), one can improve the behaviour of $b_r(\zeta)$. For instance, if $q(x)$ has rapidly decaying derivatives, then $b_r(\zeta)$ has rapidly decaying derivatives as well. The converse also holds. In particular, q is in the Schwartz class if and only if b_r is in the Schwartz class (see [12]).

Moreover, if, for a potential q satisfying (2.2-13), there exists an $\varepsilon_0 > 0$ such that $q(x) = \mathcal{O}(\exp(-2\varepsilon_0 x))$ as $x \to +\infty$, then it follows from (2.9) that for any $\varepsilon_1 > 0$ with $\varepsilon_1 < \varepsilon_0$ and $\varepsilon_1' < \operatorname{Im} \zeta_j$, $j = 1,2,\ldots,N$, the function $b_r(\zeta)$ is analytic on $0 < \operatorname{Im} \zeta < \varepsilon_1$ and continuous and bounded on $0 \leq \operatorname{Im} \zeta \leq \varepsilon_1$.

We shall call the aggregate of quantities $\{b_r(\zeta), \zeta_j, C_j^r\}$ the (right) scattering data associated with the potential q. Their importance lies in the fact, that a potential is completely determined by its scattering data.

In concluding this section, let us point out that, as in the Schrödinger case (see Chapter 2, subsection 2.1), it is usually not possible to obtain the scattering data in closed form. Also here, there is an interesting exception: for the potential $q(x) = \alpha \operatorname{sech} x$, $\alpha \in \mathbb{R}\backslash\{k+\frac{1}{2}; k \in \mathbb{Z}\}$ one can solve the scattering problem (2.1) in closed form. This is shown in Chapter 6, section 5.

3. Simplification of the inverse scattering algorithm.

Let q be any potential satisfying (2.2-13). Then q can be recovered from its scattering data $\{b_r(\zeta), \zeta_j, C_j^r\}$ by solving the inverse scattering problem.
For that purpose one defines the following functions of $s \in \mathbb{R}$

$$\Omega(s) = \Omega_d(s) + \Omega_c(s), \tag{3.1a}$$

$$\Omega_d(s) = -2i \sum_{j=1}^{N} C_j^r e^{2i\zeta_j s}, \tag{3.1b}$$

$$\Omega_c(s) = \frac{1}{\pi} \int_{-\infty}^{\infty} b_r(\zeta) e^{2i\zeta s} d\zeta. \tag{3.1c}$$

Because of (2.17-22-24a) both $\Omega_d(s)$ and $\Omega_c(s)$ are real functions. Since b_r is in $C_0 \cap L^1(\mathbb{R})$, the integral in (3.1c) converges absolutely and Ω_c belongs to $C_0 \cap L^2(\mathbb{R})$.
Next, introduce the 2×2 matrix

$$\underline{\underline{\Omega}}(s) = \begin{pmatrix} 0 & -\Omega(s) \\ \Omega(s) & 0 \end{pmatrix} \tag{3.2}$$

and consider the Gel'fand-Levitan equation (see [1], [2], [6], [12])

$$\underline{\underline{\beta}}(y;x) + \underline{\underline{\Omega}}(x+y) + \int_0^{\infty} \underline{\underline{\beta}}(z;x)\underline{\underline{\Omega}}(x+y+z)dz = 0 \tag{3.3}$$

with $y > 0$, $x \in \mathbb{R}$. In this integral equation the unknown $\underline{\underline{\beta}}(y;x)$ is a 2×2 matrix function of the variable y, whereas x is a parameter. Observe that some authors use a slightly different version of the Gel'fand-Levitan equation which can be transformed into (3.3) by a change of variables (see [6], p. 46).
In [2] it is shown that for each $x \in \mathbb{R}$ there is a unique solution $\underline{\underline{\beta}}(y;x)$ to (3.3) in $(L^2)^{2\times 2}$ $(0 < y < +\infty)$. It has the form

$$\underline{\underline{\beta}} = \begin{pmatrix} \alpha & -\beta \\ \beta & \alpha \end{pmatrix}, \tag{3.4}$$

where $\alpha(y;x)$ and $\beta(y;x)$ are real functions belonging to $C \cap L^1 \cap L^2$ $(0 < y < +\infty)$, which vanish as $y \to +\infty$. The inverse scattering problem is now solved, since the functions α and β are related to the potential q in the following way

(3.5a) $$q(x) = \beta(0^+;x)$$

(3.5b) $$\int_x^\infty q^2(s)ds = -\alpha(0^+;x), \qquad x \in \mathbb{R}.$$

Using (3.4), the matrix integral equation (3.3) can be reduced to a scalar integral equation involving only β

(3.6) $$\beta(y;x) + \Omega(x+y) + \int_0^\infty \int_0^\infty \beta(z;x)\Omega(z+s+x)\Omega(s+y+x)dsdz = 0.$$

This is the form of the Gel'fand-Levitan equation that appears in the literature (cf. [1], [12]) and is frequently used in the asymptotic analysis of nonlinear equations solvable via the Zakharov-Shabat inverse scattering formalism.

It has, in our view, a number of disadvantages. Firstly, the information about α, which was still present in (3.3), is now lost. Secondly, the equation contains a double integral which is of course harder to analyse than a single one. The third objection is of an algebraic nature: the structure of $\underline{\underline{\beta}}$, which is simply the matrix representation of a complex number, is violated.

Let us try to mend these shortcomings, starting with the last one. Set

(3.7) $$\gamma(y;x) = \alpha(y;x) + i\beta(y;x).$$

Then it is straightforward to deduce the following integral equation from (3.3-4)

(3.8) $$\gamma(y;x) + i\Omega(x+y) + i\int_0^\infty \Omega(x+y+z)\gamma(z;x)dz = 0.$$

Clearly, equation (3.8) has none of the disadvantages mentioned above. In particular, the information about α and β is obtained by taking real and imaginary parts. This yields

(3.9a) $$q(x) = \operatorname{Im} \gamma(0^+;x)$$

(3.9b) $$\int_x^\infty q^2(s)ds = -\operatorname{Re} \gamma(0^+;x).$$

Note, that the equations (3.8) and (3.3-4) are equivalent. In fact, let ϕ denote the mapping that identifies a complex number with its matrix representation in the following way

$$(3.10)\qquad \phi(\xi + i\eta) = \begin{pmatrix} \xi & -\eta \\ \eta & \xi \end{pmatrix} \qquad \xi,\ \eta \in \mathbb{R}.$$

Then (3.3-4) is the image of (3.8) under ϕ.

4. Statement of the main result.

At the moment no method is known that produces explicit solutions of the Gel'fand-Levitan equation (3.8), associated with arbitrary scattering data. However, if $b_r \equiv 0$, then the equation gets degenerated and reduces essentially to a system of N linear algebraic equations, which can be solved in explicit form by standard procedures.

If q is a potential with scattering data $\{b_r(\zeta),\zeta_j,C_j^r\}$ then the potential q_d with scattering data $\{0,\zeta_j,C_j^r\}$ is called the reflectionless part of q. The structure of the reflectionless part has been discussed by several authors (see [2], [9], [10]). In the next section we shall derive a rather elegant representation of q_d in terms of determinants.

In applications of the inverse scattering method the following situation is generic: by some procedure the scattering data of a potential are known in explicit form. The potential itself is predicted to exist by the general theory, but its explicit form is unknown. The only thing one can calculate explicitly is its reflectionless part. Thus, a natural question to ask is the following:

In which sense is the potential approximated by its reflectionless part?

Our main result is the next theorem which gives an answer to this question.

Theorem 4.1. *Let* q *be a potential in the Zakharov-Shabat system (2.1), which satisfies (2.2-13) and has the scattering data* $\{b_r(\zeta),\zeta_j,C_j^r\}$. *Let* q_d *denote the reflectionless part of* q. *Then for each* $x \in \mathbb{R}$

$$(4.1)\qquad |q(x) - q_d(x)| \le a_0^2\left(\int_0^\infty |\Omega_c(x+y)|^2\,dy + \sup_{0<y<+\infty} |\Omega_c(x+y)|\right),$$

with Ω_c *given by (3.1c) and* a_0 *the following explicit function of the bound states* ζ_j

(4.2a) $$a_0 = 1 + \sum_{p,j=1}^{N} (\text{Im}\,\zeta_p)^{-1} N_{pj},$$

(4.2b) $$N_{pj} = 2(\text{Im}\,\zeta_p)^{\frac{1}{2}}(\text{Im}\,\zeta_j)^{\frac{1}{2}} \prod_{\substack{\ell=1\\ \ell\neq p}}^{N} \left|\frac{\zeta_p-\zeta_\ell^*}{\zeta_p-\zeta_\ell}\right| \prod_{\substack{k=1\\ k\neq j}}^{N} \left|\frac{\zeta_j-\zeta_k^*}{\zeta_j-\zeta_k}\right| .$$

Herewith, $q - q_d$ *is estimated completely in terms of the scattering data. More precisely, the bound given by (4.1-2) depends only on the reflection coefficient* $b_r(\zeta)$ *and the bound states* ζ_j *and not on the normalization coefficients* C_j^r.

Corollary to theorem 4.1. *Under the conditions of theorem 4.1 we have the a priori bound*

(4.3) $$\sup_{x\in\mathbb{R}} |q(x) - q_d(x)| \leq \frac{a_0^2}{\pi} \int_{-\infty}^{\infty} (|b_r(\zeta)| + |b_r(\zeta)|^2)d\zeta.$$

Let us mention an important application of theorem 4.1. Consider a family of potentials $q(x,t)$, depending on a parameter $t \geq 0$ referred to as time. Suppose that $q(x,t)$, which is assumed sufficiently smooth and rapidly decaying for $|x| \to \infty$, satisfies the initial value problem for a suitable nonlinear evolution equation of AKNS class [2], e.g. the modified Korteweg-de Vries equation. Then the bound states ζ_j do not change with time, whereas the associated normalization coefficients and reflection coefficient change in a simple way. The estimate (4.1-2) now tells us how well the solution $q(x,t)$ is approximated by its soliton part $q_d(x,t)$. Since a_0 is invariant with time, only the behaviour of $\Omega_c(x+y;t)$ is of importance. In particular, for those nonlinear evolution equations, whose linearized version has a negative group velocity associated with it (cf. [7], or Chapter 1, Appendix C) one can construct coordinate regions of the form

(4.4) $$t \geq t_0, \quad x \geq \alpha(t)$$

with t_0 a nonnegative constant and $\alpha(t)$ a function of time characteristic for the problem, such that $\Omega_c(x+y;t)$ is small in $L^2 \cap L^\infty$ $(0 < y < +\infty)$ (see [11], or Chapter 2, section 3, for an example of such a construction). In that case, (4.1-2) shows that in the region (4.4) the solution $q(x,t)$ is given by $q_d(x,t)$ plus a small correction term, which dies out as $t \to \infty$.

We shall prove theorem 4.1 in section 6. Before doing so we introduce some notation and derive a useful lemma in section 5.

5. Auxiliary results.

From now on we shall assume that the conditions of theorem 4.1 are fulfilled.
To start with, let us give our reasoning an appropriate abstract setting. To this end we introduce the Banach space $\mathcal{B}$ of all complex-valued, continuous and bounded functions g on $(0,\infty)$, equipped with the supremum norm

$$\|g\| = \sup_{0<y<+\infty} |g(y)|.$$

Furthermore, let $\mathcal{H}$ denote the complex Hilbert space $L^2(0,\infty)$ with inner product $\langle f,g\rangle = \int_0^\infty f(y)g^*(y)dy$ and corresponding norm $\| \ \|_2$.

From section 3 we know that for each $x \in \mathbb{R}$ the functions $y \mapsto \Omega_c(x+y)$, $\Omega_d(x+y)$ belong to $\mathcal{B} \cap \mathcal{H}$.

Next, keeping $x \in \mathbb{R}$ fixed, we formally write

$$(5.1) \qquad (T_d g)(y) = \int_0^\infty \Omega_d(x+y+z)g(z)dz$$

$$(5.2) \qquad (T_c g)(y) = \int_0^\infty \Omega_c(x+y+z)g(z)dz.$$

Evidently, T_d can be considered as a mapping from $\mathcal{B}$ into $\mathcal{B}$, but equally well as a mapping from $\mathcal{H}$ into $\mathcal{H}$. On the other hand, T_c is not necessarily a mapping from $\mathcal{B}$ into $\mathcal{B}$. However, an obvious adaptation of formula (4.5.10) in [6] shows that T_c maps $\mathcal{H}$ into $\mathcal{H}$ with a norm that satisfies

$$(5.3) \qquad \|T_c\|_2 \leq \sup_{\zeta \in \mathbb{R}} |b_r(\zeta)|.$$

In view of (2.25) we do not expect this norm to be particularly small. Since both Ω_d and Ω_c are real-valued, the operators T_d and T_c are self-adjoint on $\mathcal{H}$. This fact will play a crucial role in our analysis. Actually, to such an extent that the size of $\|T_c\|_2$ is irrelevant.
In the above abstract language, the Gel'fand-Levitan equation (3.8) takes the form

$$(5.4a) \qquad (I + iT_c + iT_d)\gamma = -i\Omega$$

$$(5.4b) \qquad \Omega = \Omega_c + \Omega_d,$$

where I is the identity mapping.
A first advantage of this formulation is readily seen. Since $T_c + T_d$ is self-adjoint, the operator $I + iT_c + iT_d$ is invertible on $\mathcal{H}$ and so we know at once that (5.4) has a unique solution $\gamma \in \mathcal{H}$. Note that this fact was already mentioned in section 3, from which we recall that, moreover, $\gamma \in \mathcal{B} \cap \mathcal{H}$.

For the proof of theorem 4.1 the following lemma is basic.

Lemma 5.1. *For any value of the parameter* $x \in \mathbb{R}$, *the operator* $I+iT_d$ *is invertible on the Banach space* $\mathcal{B}$ *with inverse* $S = (I + iT_d)^{-1}$ *given by*

$$(5.5a) \qquad (Sf)(y) = f(y) - \sum_{j=1}^{N} A_j e^{2i\zeta_j y}$$

$$(5.5b) \qquad A_j = \sum_{p=1}^{N} \beta_{pj}\left(2 \int_0^\infty f(z)e^{2i\zeta_p z}\, dz\right),$$

where (β_{pj}) *is the inverse of the matrix* $A = \left([c_j^r]^{-1} e^{-2i\zeta_j x}\delta_{pj} + i(\zeta_p+\zeta_j)^{-1}\right)$.
Furthermore, the operator S *satisfies the bound*

$$(5.6) \qquad \|S\| \leq a_0, \quad x \in \mathbb{R},$$

$$(5.7a) \qquad a_0 = 1 + \sum_{p,j=1}^{N} (\mathrm{Im}\,\zeta_p)^{-1} N_{pj}$$

$$(5.7b) \qquad N_{pj} = 2(\mathrm{Im}\,\zeta_p)^{\frac{1}{2}}(\mathrm{Im}\,\zeta_j)^{\frac{1}{2}} \prod_{\substack{\ell=1\\ \ell\neq p}}^{N} \left|\frac{\zeta_p-\zeta_\ell^*}{\zeta_p-\zeta_\ell}\right| \prod_{\substack{k=1\\ k\neq j}}^{N} \left|\frac{\zeta_j-\zeta_k^*}{\zeta_j-\zeta_k}\right| .$$

Thus, $\|S\|$ *is uniformly bounded for* $x \in \mathbb{R}$ *and the bound is explicitly given in terms of the* ζ_j *only.*

Proof: Let $x \in \mathbb{R}$ be arbitrarily chosen.
We first consider T_d as an operator from the Hilbert space $\mathcal{H}$ into itself. Since T_d is self-adjoint, $I + iT_d$ is an invertible operator on $\mathcal{H}$. From the relation

$$(5.8) \qquad \|(I + iT_d)g\|_2^2 = \|g\|_2^2 + \|T_d g\|_2^2, \quad g \in \mathcal{H},$$

we obtain the following bounds for the inverse $S = (I + iT_d)^{-1}$

$$(5.9) \qquad \|S\|_2 \leq 1, \qquad \|T_d S\|_2 \leq 1.$$

Next, let us consider T_d as an operator from $\mathcal{B}$ into $\mathcal{B}$ and show that $I + iT_d$ is invertible on $\mathcal{B}$. Suppose that $(I + iT_d)g = 0$ for some $g \in \mathcal{B}$. Then $g = -iT_d g \in \mathcal{H}$ and thus g is identically zero by the preceding argument. This shows that $I + iT_d$ is one to one on $\mathcal{B}$. However, T_d is an operator of finite rank and hence compact. It follows that $I + iT_d$ is an invertible operator on the Banach space $\mathcal{B}$.

Furthermore, solving the equation

$$(5.10) \qquad (I + iT_d)g = f, \quad f,g \in \mathcal{B}$$

we find

$$(5.11) \qquad g(y) = f(y) - \sum_{j=1}^{N} A_j e^{2i\zeta_j y},$$

where the A_j satisfy

$$(5.12a) \qquad \sum_{j=1}^{N} \alpha_{pj} A_j = 2 \int_0^\infty f(z) e^{2i\zeta_p z} dz, \quad p = 1,2,\ldots,N$$

$$(5.12b) \qquad \alpha_{pj} = \alpha_j \delta_{pj} + i(\zeta_p + \zeta_j)^{-1}, \quad \alpha_j = [C_j^r]^{-1} e^{-2i\zeta_j x}.$$

Since the operator $I + iT_d$ is one to one on $\mathcal{B}$, the matrix $A = (\alpha_{pj})$ is invertible with inverse $A^{-1} = (\beta_{pj})$. We conclude that the inverse operator $S = (I + iT_d)^{-1}$ is given in explicit form by (5.5).

We shall now prove that the matrix elements β_{pj} are bounded as functions of $x \in \mathbb{R}$, where the bound is explicitly given in terms of the ζ_j by

(5.13) $$|\beta_{pj}| \leq 2(\operatorname{Im}\zeta_p)^{\frac{1}{2}}(\operatorname{Im}\zeta_j)^{\frac{1}{2}} \prod_{\substack{\ell=1\\ \ell\neq p}}^{N} \left|\frac{\zeta_p-\zeta_\ell^*}{\zeta_p-\zeta_\ell}\right| \prod_{\substack{k=1\\ k\neq j}}^{N} \left|\frac{\zeta_j-\zeta_k^*}{\zeta_j-\zeta_k}\right| \equiv N_{pj}.$$

To this end we first introduce some notation. In $\mathcal{H}$ we consider the elements e_j defined by $e_j(y) = e^{2i\zeta_j y}$. Let $\tilde{A}$ denote the Gram matrix of the vectors $e_1,e_2,\ldots,e_N$, i.e. $\tilde{A} = (\tilde{\alpha}_{pj})$, $\tilde{\alpha}_{pj} = \langle e_p,e_j\rangle = (2i(\zeta_j^* - \zeta_p))^{-1}$. Since the vectors $e_1,e_2,\ldots,e_N$ are linearly independent, it follows that $\det\tilde{A} > 0$ (see [5]). Let us write $(\tilde{A})^{-1} = (\tilde{\beta}_{pj})$ and introduce the vectors $h_p = \sum_{j=1}^{N} \tilde{\beta}_{pj}e_j$. Evidently, $\langle h_p,e_j\rangle = \delta_{pj}$ and $\langle h_p,h_j\rangle = \tilde{\beta}_{pj}$. In combination with (5.5) this gives

(5.14) $$Sh_{\sigma(p)} = h_{\sigma(p)} - 2\sum_{j=1}^{N} \beta_{pj}e_j,$$

where σ is the permutation defined in (2.18).

Using the identity $I - S = iT_dS$, we get

(5.15) $$2\beta_{pj} = \langle iT_dSh_{\sigma(p)},h_j\rangle.$$

Hence, in view of (5.9)

(5.16) $$4|\beta_{pj}|^2 \leq \|h_{\sigma(p)}\|_2^2\|h_j\|_2^2 = \tilde{\beta}_{\sigma(p)\sigma(p)}\tilde{\beta}_{jj}.$$

By direct calculation (see [5]) we obtain

(5.17) $$\tilde{\beta}_{\ell\ell} = \frac{\det\left((2i(\zeta_j^* - \zeta_p))^{-1}\right)_{p,j=1,\ p\neq\ell,\ j\neq\ell}^{N}}{\det\left((2i(\zeta_j^* - \zeta_p))^{-1}\right)_{p,j=1}^{N}} =$$

$$= 4(\operatorname{Im}\zeta_\ell) \prod_{\substack{p=1\\ p\neq\ell}}^{N} \left|\frac{\zeta_\ell-\zeta_p^*}{\zeta_\ell-\zeta_p}\right|^2 = \tilde{\beta}_{\sigma(\ell)\sigma(\ell)},$$

which completes the proof of (5.13).

By (5.5b-13) we have

(5.18) $$|A_j| \leq \|f\| \sum_{p=1}^{N} (\operatorname{Im}\zeta_p)^{-1}N_{pj},$$

from which the bound (5.6-7) for $\|S\|$ is obvious.

Corollary to lemma 5.1. *For each* $x \in \mathbb{R}$ *the equation*

(5.19) $$(I + iT_d)\gamma = -i\Omega_d$$

admits a unique solution $\gamma_d \in \mathcal{B}$ *and we have*

$$\gamma_d(y;x) = -2 \sum_{p,j=1}^{N} \beta_{pj} e^{2i\zeta_j y}. \tag{5.20}$$

Remark. Let us recall that γ_d produces the reflectionless part of the potential q through the formula

$$q_d(x) = \operatorname{Im} \gamma_d(0^+;x) = -2 \operatorname{Im} \sum_{p,j=1}^{N} \beta_{pj}. \tag{5.21}$$

Clearly, by (5.13) we have the a priori bound

$$\sup_{x\in\mathbb{R}} |q_d(x)| \le 2 \sum_{p,j=1}^{N} N_{pj}, \tag{5.22}$$

which does not involve the C_j^r but depends only on the ζ_j in a simple explicit way.

Starting from (5.21) it is easy to obtain a more elegant representation of q_d. Indeed, introducing the matrices

$$V = \left(\delta_{pj} + i(\zeta_p+\zeta_j)^{-1} C_j^r e^{i(\zeta_p+\zeta_j)x}\right), \quad D_1 = \left(e^{i\zeta_j x}\delta_{pj}\right) \tag{5.23a}$$

$$D_2 = \left(C_j^r e^{i\zeta_j x}\delta_{pj}\right), \qquad E = (\varepsilon_{pj}),\ \varepsilon_{pj} = 1 \tag{5.23b}$$

we find that

$$\sum_{p,j=1}^{N} \beta_{pj} = \operatorname{Tr}(EA^{-1}) = \operatorname{Tr}(ED_2V^{-1}D_1) = \operatorname{Tr}(D_1ED_2V^{-1}), \tag{5.24}$$

where we used the fact that $V = D_1AD_2$, as well as the invariance of the trace under cyclic permutation. Clearly, $D_1ED_2 = -\frac{d}{dx}V$. Therefore, using the familiar formula (cf. [4], p. 28, (7.17))

$$\frac{d}{dx}(\det V) = (\det V)\operatorname{Tr}\left(\frac{dV}{dx}V^{-1}\right) \tag{5.25}$$

we arrive at

$$\gamma_d(0^+;x) = 2\frac{d}{dx}\log\det V. \tag{5.26}$$

An alternative derivation of formula (5.26) is given in [9], pp. 103-105. Transforming back one gets

(5.27) $$\gamma_d(0^+;x) = 2\frac{d}{dx}\log[\det(D_1D_2)\det A]$$

$$= 4i\sum_{p=1}^{N}\zeta_p + 2\frac{d}{dx}\log\det A.$$

Note that $\sum_{p=1}^{N}\zeta_p$ is purely imaginary. We conclude that

(5.28a) $$q_d(x) = 2\mathrm{Im}\frac{d}{dx}\log\det A$$

or equivalently

(5.28b) $$q_d(x) = 2\ \mathrm{Im}\ \mathrm{Tr}(\frac{dA}{dx}A^{-1}) =$$

$$= \mathrm{Im}\sum_{p=1}^{N}(-4i\zeta_p\alpha_p)\beta_{pp} \text{ with } \alpha_p = [c_p^r]^{-1}e^{-2i\zeta_p x}.$$

Observe how this simplifies (5.21). For time-dependent potentials the representations (5.28) form a good starting point for the asymptotic analysis of $q_d(x,t)$ as $t \to +\infty$ (cf. [6], 2nd ed., Appendix A1).

6. Proof of theorem 4.1.

After the preparatory work done in section 5 the proof of theorem 4.1, which we present in this section, is comparatively easy.

Let $x \in \mathbb{R}$ be arbitrarily fixed. As a first step, let us write the solution γ of (5.4) in the following form

(6.1) $$\gamma = \gamma_d + \gamma_c, \text{ with}$$

(6.2) $$\gamma_d = -iS\Omega_d.$$

By (3.9a) and (5.21) we plainly have

(6.3) $$q(x) - q_d(x) = \mathrm{Im}\ \gamma_c(0^+;x).$$

From the previous section it is clear that both γ and γ_d belong to $\mathcal{B} \cap \mathcal{H}$. Hence, we already know that $\gamma_c \in \mathcal{B} \cap \mathcal{H}$. It remains to find a concrete estimate of γ_c in the supremum norm. For that purpose we insert the decomposition (6.1) into (5.4), thereby obtaining

(6.4) $$(I + iT_c + iT_d)\gamma_c = -iT_c\gamma_d - i\Omega_c.$$

Consider (6.4) as an equation in the Hilbert space $\mathcal{H}$. Since $T_c + T_d$ is self-adjoint, the operator $I + iT_c + iT_d$ is invertible on $\mathcal{H}$. Furthermore, the relation (5.8) holds with T_d replaced by $T_c + T_d$. Consequently, (6.4) has a unique solution $\gamma_c \in \mathcal{H}$ satisfying

(6.5) $$\|\gamma_c\|_2 \leq \|T_c\gamma_d\|_2 + \|\Omega_c\|_2.$$

An application of the generalized Minkowski inequality (see [8], p. 148) gives us

(6.6) $$\|T_c\gamma_d\|_2 \leq \int_0^\infty |\gamma_d(z;x)|\left(\int_0^\infty |\Omega_c(x+y+z)|^2 dy\right)^{\frac{1}{2}} dz.$$

Hence

(6.7) $$\|T_c\gamma_d\|_2 \leq \|\Omega_c\|_2\|\gamma_d\|_1,$$

where by (5.13-20)

(6.8) $$\|\gamma_d\|_1 \equiv \int_0^\infty |\gamma_d(z;x)|dz \leq \sum_{p,j=1}^{N} (\mathrm{Im}\ \zeta_p)^{-1} N_{pj}.$$

We conclude that

(6.9) $$\|\gamma_c\|_2 \leq \|\Omega_c\|_2(1 + \|\gamma_d\|_1) \leq a_0\|\Omega_c\|_2$$

with a_0 given by (4.2a). The trick is now to rewrite equation (6.4) as

(6.10) $$(I + iT_d)\gamma_c = -iT_c\gamma_c - iT_c\gamma_d - i\Omega_c$$

and to realize that the a priori estimate (6.9) paves the way to estimate the right hand side of (6.10) in the supremum norm. In fact, we have by Schwarz' inequality

(6.11) $$\|T_c\gamma_c\| \leq \sup_{0<y<+\infty} \left(\int_0^\infty |\Omega_c(x+y+z)|^2 dz\right)^{\frac{1}{2}}\left(\int_0^\infty |\gamma_c(z;x)|^2 dz\right)^{\frac{1}{2}}$$

$$\leq \|\Omega_c\|_2\|\gamma_c\|_2 \leq \|\Omega_c\|_2^2(1 + \|\gamma_d\|_1).$$

Moreover, invoking again the generalized Minkowski inequality, one gets

(6.12) $$\|T_c\gamma_d\| \leq \|\Omega_c\|\|\gamma_d\|_1.$$

Together, (6.11) and (6.12) lead to the estimate

$$(6.13)\qquad \|-iT_c\gamma_c - iT_c\gamma_d - i\Omega_c\| \leq \left(\|\Omega_c\| + \|\Omega_c\|_2^2\right)\left(1 + \|\gamma_d\|_1\right).$$

Applying lemma 5.1 we obtain from (6.10-13) the following estimate for γ_c in the supremum norm

$$(6.14)\qquad \|\gamma_c\| \leq a_0^2\left(\|\Omega_c\| + \|\Omega_c\|_2^2\right).$$

The desired estimate (4.1-2) is now a direct consequence of (6.3-14) and the obvious fact

$$(6.15)\qquad |\mathrm{Im}\ \gamma_c(0^+;x)| \leq \sup_{0<y<+\infty} |\gamma_c(y;x)| = \|\gamma_c\|.$$

Herewith the proof of theorem 4.1 is completed.

Remark. Note that we have proven more, since by (6.1-14) and (3.9b) the estimate (4.1-2) is still valid if one replaces the left hand side of (4.1) by

$$(6.16)\qquad \max\left[|q(x) - q_d(x)|,\ \left|\int_x^\infty (q^2(s) - q_d^2(s))ds\right|\right].$$

References

[1] M.J. Ablowitz, Lectures on the inverse scattering transform, Stud. Appl. Math 58 (1978), 17-94.

[2] M.J. Ablowitz, D.J. Kaup, A.C. Newell and H. Segur, The inverse scattering transform-Fourier analysis for nonlinear problems, Stud. Appl. Math. 53 (1974), 249-315.

[3] M.J. Ablowitz and H. Segur, Solitons and the Inverse Scattering Transform, Philadelphia, SIAM, 1981.

[4] E.A. Coddington and N. Levinson, Theory of Ordinary Differential Equations, Mc Graw-Hill, 1955.

[5] P.J. Davis, Interpolation and Approximation, Dover, New York, 1963.

[6] W. Eckhaus and A. van Harten, The Inverse Scattering Transformation and the Theory of Solitons, North-Holland Mathematics Studies 50, 1981 (2nd ed. 1983).

[7] W. Eckhaus and P. Schuur, The emergence of solitons of the Korteweg-de Vries equation from arbitrary initial conditions, Math. Meth. in the Appl. Sci. 5 (1983), 97-116.

[8] G.H. Hardy, J.E. Littlewood and G. Pólya, Inequalities, 2nd. ed., Cambridge, 1952.

[9] G.L. Lamb Jr., Elements of Soliton Theory, Wiley-Interscience, 1980.

[10] M. Ohmiya, On the generalized soliton solutions of the modified Korteweg-de Vries equation, Osaka J. Math. 11 (1974), 61-71.

[11] P. Schuur, Asymptotic estimates of solutions of the Korteweg-de Vries equation on right half lines slowly moving to the left, Preprint 330 Mathematical Institute Utrecht (1984).

[12] S. Tanaka, Non-linear Schrödinger equation and modified Korteweg-de Vries equation; construction of solutions in terms of scattering data, Publ. R.I.M.S. Kyoto Univ. 10 (1975), 329-357.

[13] S. Tanaka, Some remarks on the modified Korteweg-de Vries equations, Publ. R.I.M.S. Kyoto Univ. 8 (1972/73), 429-437.

[14] V.E. Zakharov and A.B. Shabat, Exact theory of two-dimensional self-focusing and one-dimensional self-modulation of waves in non-linear media, Soviet Phys. JETP (1972), 62-69.

CHAPTER FIVE

DECOMPOSITION AND ESTIMATES OF SOLUTIONS OF THE MODIFIED KORTEWEG-DE VRIES EQUATION ON RIGHT HALF LINES SLOWLY MOVING LEFTWARD

We consider the modified Korteweg-de Vries equation $q_t + 6q^2 q_x + q_{xxx} = 0$ with arbitrary real initial conditions $q(x,0) = q_0(x)$, sufficiently smooth and rapidly decaying as $|x| \to \infty$, such that q_0 is a bona fide potential in the Zakharov-Shabat scattering problem. Using the method of the inverse scattering transformation we analyse the behaviour of the solution $q(x,t)$ in all coordinate regions of the form $T = (3t)^{1/3} > 0$, $x \geqq - \mu - \nu T$, where μ and ν are arbitrary nonnegative constants. We derive explicit x and t dependent bounds for the non-reflectionless part of $q(x,t)$. If all bound states are purely imaginary these bounds make it possible to derive a convergence result, clearly displaying the emergence of solitons. Furthermore, if the reflectionless part of the solution is confined to $x > 0$ as $t \to \infty$, then the bounds help us to establish some interesting energy decomposition formulae.

1. Introduction.

We focus our attention on the modified Korteweg-de Vries (mKdV) problem

(1.1a) $q_t + 6q^2 q_x + q_{xxx} = 0, \quad -\infty < x < +\infty, \quad t > 0$

(1.1b) $q(x,0) = q_0(x),$

where the initial function $q_0(x)$ is an arbitrary real function on $\mathbb{R}$, such that

(1.2a) $q_0(x)$ satisfies the hypotheses (2.2-13) in [8] (i.e. Chapter 4) and is therefore a bona fide potential in the Zakharov-Shabat scattering problem.

(1.2b) $q_0(x)$ is sufficiently smooth and (along with a number of its derivatives) decays sufficiently rapidly for $|x| \to \infty$:

(i) for the whole of the Zakharov-Shabat inverse scattering method [1] to work,

(ii) to guarantee certain regularity and decay properties of the right reflection coefficient to be stated further on.

Note that the method of L^2-energy estimates yields uniqueness of solutions of (1.1) within the class of functions which, together with a sufficient number of derivatives vanish for $|x| \to \infty$ (cf. [3]). Let us refer to this class as the "modified Lax-class" (after [4]).

In [10] it is shown by an inverse scattering analysis that condition (1.2a) and a special case of condition (1.2b) (namely q_0 in Schwartz space) guarantee the existence of a real function $q(x,t)$, continuous on $\mathbb{R}\times[0,\infty)$, which satisfies (1.1) in the classical sense. Whenever, in this chapter, we speak of "the solution" of (1.1) we shall refer to the solution obtained by inverse scattering (unique within the modified Lax-class).

Let us recall that by the inverse scattering method the solution $q(x,t)$ of (1.1) is obtained in the following way. First one computes the (right) scattering data $\{b_r(\zeta),\zeta_j,C_j^r\}$ associated with $q_0(x)$. For their definition and properties we refer to Chapter 4. Next one puts (see [1], [10])

(1.3a) $C_j^r(t) = C_j^r \exp\{8i\zeta_j^3 t\}$, $j = 1,2,\ldots,N$

(1.3b) $b_r(\zeta,t) = b_r(\zeta)\exp\{8i\zeta^3 t\}$, $-\infty < \zeta < +\infty$.

Then by the solvability of the inverse scattering problem, there exists for each $t > 0$ a smooth potential $q(x,t)$ satisfying the hypotheses (2.2-13) in Chapter 4 and having $\{b_r(\zeta,t),\zeta_j,C_j^r(t)\}$ as its scattering data ([10]). The function $q(x,t)$ is the unique solution of the mKdV initial value problem (1.1).

Explicit solutions by the above procedure have only been obtained for $b_r \equiv 0$. The solution $q_d(x,t)$ of the mKdV equation with scattering data $\{0,\zeta_j,C_j^r(t)\}$ is called the reflectionless part of $q(x,t)$.

The long-time behaviour of the solution $q(x,t)$ of (1.1) in the absence of solitons is discussed in [2], [9].
In [2] the existence is claimed of three distinct asymptotic regions

I. $x \geq \tilde{O}(t)$ II. $|x| \leq \tilde{O}(t^{1/3})$

III. $-x \geq \tilde{O}(t)$.

Here $\tilde{O}$ denotes positive proportionality. Within each region, the solution $q(x,t)$ has an asymptotic expansion, characteristic for that region. However, as stated in [2], p. 68 proofs are yet to be given. Apparently, the asymptotic structure of the solitonless solution is simpler for mKdV than for KdV, where four asymptotic regions were found.

When solitons are present the degrees of complexity are reversed. The reason is the asymptotic structure of the reflectionless part $q_d(x,t)$. First of all, depending on the location of the bound states ζ_j, it may happen that solitons do not separate out as $t \to +\infty$. Furthermore, if they do separate out, one can expect both sech-shaped solitons as well as breathers (see the discussion in section 2). The sech-shaped solitons move nicely to the right, but the breathers are unpredictable: depending on the ζ_j, the breather envelope may move to the right, to the left, or even be at rest. Moreover, if one relaxes condition (1.2a) to Chapter 4, (2.2-13a) and uses the inverse scattering formalism developed by Tanaka in [10], the asymptotic structure of the associated reflectionless part of $q(x,t)$ is even more complicated, since now multiple-pole solutions can occur, as calculated in [11]. In the sequel, however, we shall stick to

(2.2-13). In summary, only under additional restrictions on the location of the bound states can we expect $q_d(x,t)$ to decompose into a finite number of solitons moving to the right.

However, we prefer to consider the general situation. In that case, the only thing about $q(x,t)$ one can expect (in view of the negative group velocity associated with the linearized version of (1.1)) is the leftward motion of the dispersive wavetrain.
Thus we are confronted with the question: Confining ourselves to the regions I and II how well is the solution $q(x,t)$ approximated by its reflectionless part $q_d(x,t)$?

In this chapter we give a complete answer to this question by analysing the behaviour of the function $q(x,t) - q_d(x,t)$ in all coordinate regions of the form

$$(1.4) \qquad T = (3t)^{1/3} > 0, \quad x \geq -\alpha, \quad \alpha = \mu + \nu T,$$

where μ and ν are arbitrary nonnegative constants. Plainly, (1.4) covers all of the regions I and II. It is proven that

$$(1.5) \qquad \sup_{x \geq -\alpha} |q(x,t) - q_d(x,t)| = O(t^{-1/3}) \quad \text{as } t \to \infty.$$

If all bound states are purely imaginary, then (1.5) leads to a convergence result (see (5.5)) which clearly displays the emergence of solitons. Moreover, we construct several explicit x and t dependent bounds for $q(x,t) - q_d(x,t)$ valid in the region (1.4). In the special case that the reflectionless part is confined to $x > 0$ as $t \to \infty$, we derive the energy decomposition formulae

$$(1.6a) \qquad \int_{-\alpha}^{\infty} q^2(x,t)dx = 4 \sum_{p=1}^{N} \operatorname{Im} \zeta_p + O(t^{-1/3}) \qquad \text{as } t \to \infty$$

$$(1.6b) \qquad \int_{-\infty}^{-\alpha} q^2(x,t)dx = \frac{2}{\pi} \int_0^{\infty} \log(1 + |b_r(\zeta)|^2)d\zeta + O(t^{-1/3}) \quad \text{as } t \to \infty.$$

The results obtained in this chapter are of a similar nature as those obtained for KdV in [7], i.e. Chapter 2 in this volume. However, there are differences. For instance, in the KdV case we could improve (1.5) in the absence of solitons (i.e. $q_d \equiv 0$). The analysis in this chapter indicates that no such improvement is likely for mKdV. Furthermore, the

KdV estimates in Chapter 2 are only valid for $0 \leq \nu < \nu_c$ where ν_c is some fixed number connected with properties of the Airy function, and for $t \geq t_{cd}$ where t_{cd} is some critical time. The mKdV estimates obtained here are valid for any value of $\nu \geq 0$ and for all $t > 0$. A third difference is found in the structure of the energy decomposition formulae (1.6) versus Chapter 2, (7.25). Finally, we refer to the remarks about the asymptotic structure of the reflectionless part made above.

The organization of this chapter is as follows.
In section 2 we isolate certain properties of the reflectionless part $q_d(x,t)$. In section 3 we recall two essential results obtained previously in Chapter 4 and Chapter 2. The first is a general theorem in which $q(x,t) - q_d(x,t)$ is estimated in terms of $\Omega_c(x+y;t)$. The second is a lemma revealing the structure of $\Omega_c(x+y;t)$. In section 4 we apply this lemma to estimate $\Omega_c(x+y;t)$. Then in section 5 the estimates of $\Omega_c(x+y;t)$ and the theorem are combined to give estimates of $q(x,t) - q_d(x,t)$.

2. Some comments on the asymptotic structure of the reflectionless part.

The reflectionless part of $q(x,t)$ is given in explicit form (see Chapter 4, (5.28)) by

(2.1) $$q_d(x,t) = 2 \operatorname{Im} \frac{\partial}{\partial x} \log \det A$$

where $A = (\alpha_{pj})$ denotes the N×N matrix with elements

(2.2) $$\alpha_{pj} = [C_j^r(t)]^{-1} e^{-2i\zeta_j x} \delta_{pj} + i(\zeta_p + \zeta_j)^{-1}.$$

If all the bound states are purely imaginary, say $\zeta_j = i\eta_j$, $0 < \eta_N < \ldots < \eta_2 < \eta_1$, then the asymptotic structure of $q_d(x,t)$ is relatively simple. In that case, corresponding to $M = 0$ in Chapter 4, (2.18), the normalization coefficients are purely imaginary as well, say $C_j^r = i\mu_j$, $\mu_j \in \mathbb{R}\backslash\{0\}$. It is shown in [6], that as t approaches infinity the reflectionless part of $q(x,t)$ decomposes into N solitons uniformly with respect to x on $\mathbb{R}$. Specifically

$$\text{(2.3a)} \qquad \lim_{t\to\infty} \sup_{x\in\mathbb{R}} \left| q_d(x,t) - \sum_{p=1}^{N} \left(-2\eta_p \operatorname{sgn}(\mu_p) \operatorname{sech}[2\eta_p(x - x_p^+ - 4\eta_p^2 t)] \right) \right| = 0,$$

$$\text{(2.3b)} \qquad x_p^+ = \frac{1}{2\eta_p} \log\left\{ \frac{|\mu_p|}{2\eta_p} \prod_{\ell=1}^{p-1} \left(\frac{\eta_\ell - \eta_p}{\eta_\ell + \eta_p} \right)^2 \right\}.$$

Note how closely this resembles the corresponding formula Chapter 2, (5.21) for the KdV case.

If M > 0 in Chapter 4, (2.18), then the structure of $q_d(x,t)$ is more complicated. Let us consider the simplest case: $N = 2$, $\zeta_1 = \xi + i\eta$, $\zeta_2 = -\xi + i\eta = -\zeta_1^*$, with ξ, $\eta > 0$. Then $C_1^r = \lambda + i\mu$ and $C_2^r = -\lambda + i\mu$, where λ and μ are real constants that do not vanish simultaneously. Using (2.1) one gets (see [5], [11])

$$\text{(2.4)} \qquad q_d(x,t) = 4\eta \operatorname{sech} \Psi \left[\frac{\sin \Phi + (\eta/\xi) \cos \Phi \tanh \Psi}{1 + (\eta/\xi)^2 \cos^2 \Phi \operatorname{sech}^2 \Psi} \right],$$

with

$$\text{(2.5a)} \qquad \Phi = 2\xi x + 8\xi(\xi^2 - 3\eta^2)t + \phi$$

$$\text{(2.5b)} \qquad \Psi = 2\eta x + 8\eta(3\xi^2 - \eta^2)t + \psi,$$

where the constants ϕ and ψ satisfy

$$\text{(2.6a)} \qquad \exp(-\psi) = \tfrac{1}{2}(\xi/\eta)(\lambda^2 + \mu^2)^{\frac{1}{2}}(\xi^2 + \eta^2)^{-\frac{1}{2}}$$

$$\text{(2.6b)} \qquad \sin \phi = (-\lambda\eta + \mu\xi)(\lambda^2 + \mu^2)^{-\frac{1}{2}}(\xi^2 + \eta^2)^{-\frac{1}{2}}$$

$$\text{(2.6c)} \qquad \cos \phi = (\lambda\xi + \mu\eta)(\lambda^2 + \mu^2)^{-\frac{1}{2}}(\xi^2 + \eta^2)^{-\frac{1}{2}}.$$

Thus $q_d(x,t)$ has the structure of an oscillating function that is modulated by an envelope having the shape of a hyperbolic secant. The envelope and phase velocities are found from (2.5) to be $v_e = 4(\eta^2 - 3\xi^2)$ and $v_{ph} = 4(3\eta^2 - \xi^2)$, respectively. Because of the undulations in its profile, this solution is usually referred to as a breather. Note that the sign of v_e is undetermined. Hence, the breather envelope may propagate to the right, to the left or be at rest.

If M > 0 in Chapter 4, (2.18) and N > 2, then, generically, $q_d(x,t)$ will decompose into M breathers and (N − 2M) sech-shaped solitons.

However, it is easy to construct examples (e.g. $N = 3$, $\zeta_1 = \xi + i\eta$, $\eta^2 > 3\xi^2$, $\zeta_3 = i(\eta^2 - 3\xi^2)^{\frac{1}{2}}$), in which no complete decomposition into breathers and sech-shaped solitons takes place.

Looking at the envelope velocity v_e, one would expect that for large t the function $q_d(x,t)$ will be concentrated on the positive x-axis, provided that the breather bound states $\zeta_p = \xi_p + i\eta_p$ satisfy

$$(2.7) \qquad \eta_p^2 - 3\xi_p^2 > 0, \quad p = 1,2,\ldots,2M.$$

Let us confirm this mathematically. By Chapter 4, (5.28b) we have

$$(2.8) \qquad q_d(x,t) = \operatorname{Im} \sum_{p=1}^{N} \left(-4i\zeta_p[C_p^r(t)]^{-1} e^{-2i\zeta_p x}\right)\beta_{pp},$$

where $B = (\beta_{pj})$ is the inverse of the matrix A given by (2.2). Hence, using Chapter 4, (5.13), one gets

$$(2.9a) \qquad |q_d(x,t)| \le \sum_{p=1}^{N} 4|\zeta_p|N_{pp}|C_p^r|^{-1}\exp\{-\psi_p(x,t)\}$$

$$(2.9b) \qquad \psi_p(x,t) = 8t(\operatorname{Im}\zeta_p)((\operatorname{Im}\zeta_p)^2 - 3(\operatorname{Re}\zeta_p)^2) - 2(\operatorname{Im}\zeta_p)x,$$

where the constants N_{pj}, introduced in Chapter 4, (5.7b) reappear in this paper in (3.4b). Now, suppose (2.7) holds, then by (2.9)

$$(2.10a) \qquad \sup_{x<0} |q_d(x,t)| = O(e^{-\omega t}) \qquad \text{as } t \to \infty$$

$$(2.10b) \qquad \int_{-\infty}^{0} q_d^2(x,t)dx = O(e^{-2\omega t}) \qquad \text{as } t \to \infty,$$

where ω is the positive constant defined by

$$(2.11) \qquad \omega = \min\left\{8(\operatorname{Im}\zeta_p)((\operatorname{Im}\zeta_p)^2 - 3(\operatorname{Re}\zeta_p)^2); \quad p = 1,2,\ldots,N\right\}.$$

Thus, indeed, (2.7) implies that, as time goes on, $q_d(x,t)$ is confined to the positive x-axis.

With regard to (2.10b), recall that (see [2])

$$(2.12) \qquad \int_{-\infty}^{\infty} q_d^2(x,t)dx = 4\sum_{p=1}^{N} \operatorname{Im}\zeta_p.$$

3. Two useful results obtained previously.

To start with let us benefit by the work we have done in Chapter 4. Since for each $t > 0$ the solution $q(x,t)$ of (1.1) satisfies the hypotheses (2.2-13) in Chapter 4, required for a bona fide potential in the Zakharov-Shabat scattering problem, we immediately have the following result, which is established in Chapter 4 in the form of theorem 4.1 and its corollary.

Theorem 3.1. *Let $q(x,t)$ be the solution of the modified Korteweg-de Vries problem*

$$(3.1) \qquad \begin{cases} q_t + 6q^2q_x + q_{xxx} = 0, & -\infty < x < +\infty, \quad t > 0 \\ q(x,0) = q_0(x), & \end{cases}$$

where the initial function $q_0(x)$ is an arbitrary real function on $\mathbb{R}$, satisfying (1.2a-b(i)). Let $\{b_r(\zeta),\zeta_j,C_j^r\}$ be the scattering data associated with $q_0(x)$. Then for each $x \in \mathbb{R}$ and $t > 0$ one has

$$(3.2) \qquad |q(x,t) - q_d(x,t)| \leq a_0^2\left(\int_0^\infty |\Omega_c(x+y;t)|^2 dy + \sup_{0<y<+\infty} |\Omega_c(x+y;t)|\right),$$

where $q_d(x,t)$ is the reflectionless part of $q(x,t)$ given by (2.1) and

$$(3.3) \qquad \Omega_c(s;t) = \frac{1}{\pi}\int_{-\infty}^{\infty} b_r(\zeta)e^{2i\zeta s+8i\zeta^3 t}d\zeta, \quad s \in \mathbb{R}$$

$$(3.4a) \qquad a_0 = 1 + \sum_{p,j=1}^{N} (\operatorname{Im}\zeta_p)^{-1} N_{pj},$$

$$(3.4b) \qquad N_{pj} = 2(\operatorname{Im}\zeta_p)^{\frac{1}{2}}(\operatorname{Im}\zeta_j)^{\frac{1}{2}} \prod_{\substack{\ell=1\\ \ell\neq p}}^{N} \left|\frac{\zeta_p-\zeta_\ell^*}{\zeta_p-\zeta_\ell}\right| \prod_{\substack{k=1\\ k\neq j}}^{N} \left|\frac{\zeta_j-\zeta_k^*}{\zeta_j-\zeta_k}\right| .$$

Furthermore, the following a priori bound is valid

$$(3.5) \qquad \sup_{(x,t)\in\mathbb{R}\times[0,\infty)} |q(x,t) - q_d(x,t)| \leq \frac{a_0^2}{\pi}\int_{-\infty}^{\infty}\left(|b_r(\zeta)| + |b_r(\zeta)|^2\right)d\zeta.$$

Note that a_0 is invariant with time. Hence, to get an idea of the magnitude of $q(x,t) - q_d(x,t)$ in the region (1.4), only the behaviour of $\Omega_c(x+y;t)$ is important. Fortunately, we can now once again benefit from our previous work. In fact, a direct quotation of Chapter 2, theorem 3.3 gives us

Lemma 3.2. *In the situation of theorem 3.1, assume that the right reflection coefficient* $b_r(\zeta)$ *satisfies*

(3.6) b_r *is of class* $C^2(\mathbb{R})$ *and the derivatives* $b_r^{(j)}(\zeta)$, $j = 0,1,2$ *are bounded on* $\mathbb{R}$.

Let $y > 0$, $x \in \mathbb{R}$, $t > 0$. *Furthermore, let* μ, ν *denote arbitrary non-negative constants. Put*

$$w = x+y+\mu, \qquad b(\zeta,\mu) = b_r(\zeta)e^{-2i\zeta\mu}, \qquad b^{(j)} = \left(\frac{\partial}{\partial\zeta}\right)^j b \tag{3.7a}$$

$$T = (3t)^{1/3}, \qquad Z = w(3t)^{-1/3}. \tag{3.7b}$$

Then one has the representation

$$\Omega_c(x+y;t) = T^{-1}b_r(0)Ai(Z) - \tfrac{1}{2}iT^{-2}b^{(1)}(0,\mu)Ai^{(1)}(Z) + R(Z,T,\mu), \tag{3.8}$$

$$|R(Z,T,\mu)| \leq T^{-3}\|b^{(2)}\|_\infty \frac{7}{32} Z^{-3/2} \qquad \textit{for } T > 0,\ Z > 0, \tag{3.9a}$$

$$|R(Z,T,\mu)| \leq T^{-3}\|b^{(2)}\|_\infty \frac{1}{4} C(\nu) \qquad \textit{for } T > 0,\ Z \geq -\nu, \tag{3.9b}$$

with $C(\nu)$ *as in Chapter 2, (3.51b) and with* $Ai(\eta)$ *the Airy function Chapter 2, (3.3).*

Note that the first term in the representation (3.8) is directly connected with the initial function $q_0(x)$, since by Chapter 4, (2.25)

$$b_r(0) = -\tan\left[\int_{-\infty}^{\infty} q_0(s)ds\right]. \tag{3.10}$$

4. Estimates of $\Omega_c(x+y;t)$.

Let us assume that the requirements of lemma 3.2 are fulfilled. Then it is readily verified that in the parameter region

(4.1) $T = (3t)^{1/3} \geq 1$, $x \geq -\alpha$, $\alpha = \mu + \nu T$, where μ and ν are nonnegative constants,

the following estimates hold

(4.2a) $$\sup_{0<y<+\infty} |\Omega_c(x+y;t)| \leq \gamma T^{-1}$$

(4.2b) $$\int_0^{\infty} |\Omega_c(x+y;t)|dy \leq |b_r(0)|\left(\frac{1}{3} + \int_{-\nu}^{0} |Ai(\eta)|d\eta\right) + \gamma T^{-1}$$

where the constant γ is given by Chapter 2, (3.66a), with N_2 replaced by $\frac{1}{2}\|b^{(2)}\|_\infty$.

In addition to the bounds (4.2), which depend only on t, lemma 3.2 also gives us useful bounds containing both x and t. In particular, fixing $\theta \in (0,1)$ in Chapter 2, (3.7), we obtain in the region (4.1)

(4.3) $$\sup_{0<y<+\infty} |\Omega_c(x+y;t)| \leq \gamma_\theta T^{-1}\exp\left[-\frac{2}{3}\,\theta\left(\frac{x+\alpha}{T}\right)^{3/2}\right] + \rho T^{-3}\left(1 + \frac{x+\alpha}{T}\right)^{-3/2}$$

where ρ denotes the constant Chapter 2, (3.56a) with N_2 replaced by $\frac{1}{2}\|b^{(2)}\|_\infty$ and γ_θ is given by Chapter 2, (3.74).

If $q_0(x)$ enjoys the special property

(4.4) $$\int_{-\infty}^{\infty} q_0(s)ds = k\pi, \quad k \in \mathbb{Z},$$

then, in view of (3.10), the estimates (4.2-3) can be improved. For instance, by lemma 3.2 one has in the parameter region (4.1)

(4.5) $$\sup_{0<y<+\infty} |\Omega_c(x+y;t)| \leq \gamma T^{-2},$$

with γ as in (4.2). Moreover, (4.3) holds with the factor T^{-1} in front of the exponential function on the right replaced by T^{-2}. Of course, further simplifications of this type occur when also $b_r^{(1)}(0) = 0$. If, as usual, (4.4) is not fulfilled, one may simplify the representation (3.8) by working with $\mu = b_r^{(1)}(0)/(2ib_r(0))$, thereby removing the derivative of the Airy function.

Though in the present discussion we keep $\mu \geq 0$ arbitrary, our choice of the bound (4.3) is motivated by this property.

Let us point out that (3.6) is only a weak condition on the reflection coefficient, which can easily be fulfilled. Actually, as noted in Chapter 4, if $q_0(x)$ has rapidly decreasing derivatives then the same

is true for $b_r(\zeta)$.
In particular, if q_0 is in the Schwartz class, then so is b_r. In that case, there is not a shadow of a doubt that (3.6) holds.

5. Estimates of $q(x,t) - q_d(x,t)$.

Combining theorem 3.1 with the estimates of $\Omega_c(x+y;t)$ obtained in section 4 we arrive at the main result of this chapter, which can be stated as follows

Theorem 5.1. *Let* $q(x,t)$ *be the solution of the modified Korteweg-de Vries problem*

$$\text{(5.1)} \qquad \begin{cases} q_t + 6q^2 q_x + q_{xxx} = 0, & -\infty < x < +\infty, \quad t > 0 \\ q(x,0) = q_0(x), & \end{cases}$$

where the initial function $q_0(x)$ *is an arbitrary real function on* $\mathbb{R}$, *satisfying (1.2) in such a way that (3.6) is fulfilled. Let* $\{b_r(\zeta),\zeta_j,c_j^r\}$ *be the scattering data associated with* $q_0(x)$. *Then for each* $x \in \mathbb{R}$ *and* $t > 0$ *one has*

$$\text{(5.2)} \qquad |q(x,t) - q_d(x,t)| \le a_0^2\left(\int_0^\infty |\Omega_c(x+y;t)|^2 dy + \sup_{0<y<+\infty} |\Omega_c(x+y;t)|\right),$$

with $q_d(x,t)$ *the reflectionless part (2.1) of* $q(x,t)$, a_0 *the constant given by (3.4) and* Ω_c *the function introduced in (3.3).*

Next, let μ, ν *be arbitrary nonnegative constants. Put* $\alpha = \mu + \nu T$, $T = (3t)^{1/3}$. *Then the following estimate holds*

$$\text{(5.3a)} \qquad \sup_{x\ge-\alpha} |q(x,t) - q_d(x,t)| \le A \qquad \textit{for } t > 0$$

$$\text{(5.3b)} \qquad \sup_{x\ge-\alpha} |q(x,t) - q_d(x,t)| \le \tilde{\gamma} T^{-1} \qquad \textit{for } t \ge \frac{1}{3}$$

where

$$\text{(5.4a)} \qquad A = \frac{a_0^2}{\pi} \int_{-\infty}^{\infty} (|b_r(\zeta)| + |b_r(\zeta)|^2)d\zeta$$

$$(5.4b) \qquad \tilde{\gamma} = a_0^2\gamma\left[1 + \gamma + |b_r(0)|\left(\frac{1}{3} + \int_{-\nu}^{0} |Ai(\eta)|d\eta\right)\right]$$

with the constant γ *as in (4.2).*

Clearly, if in addition q_0 satisfies (4.4), then we can improve the estimate (5.3-4) since T^{-1} can be replaced by T^{-2}. Although similar remarks apply to the estimates below, they are omitted to avoid interference with the reasoning.

Note the similarity between (5.3-4) and the corresponding estimate Chapter 2, (7.16) in the KdV case. There are, however, two important differences. Firstly, the estimate (5.3-4) holds for all $t > 0$, whereas Chapter 2, (7.16) is only valid for $t \geq t_{cd}$, where t_{cd} is a certain critical time. Secondly, the results of Chapter 2, theorem 7.1, including (7.16), hold only for restricted ν-values, whereas theorem 5.1 is valid for any value of $\nu \geq 0$.

Let us discuss some consequences of theorem 5.1.

Firstly, by combining (5.3-4) with (2.3) it is found that, if all bound states are purely imaginary, the solution $q(x,t)$ of (5.1) splits up into N solitons as $t \to \infty$ in the following way

Corollary to theorem 5.1. *Suppose that* $q_0(x)$ *has the scattering data* $\{b_r(\zeta),\zeta_j,C_j^r\}$ *with* $\zeta_j = i\eta_j$, $0 < \eta_N < \ldots < \eta_2 < \eta_1$ *and* $C_j^r = i\mu_j$, $\mu_j \in \mathbb{R}\setminus\{0\}$.

Then the solution of (5.1) satisfies

$$(5.5) \qquad \lim_{t\to\infty} \sup_{x\geq -\nu T} \left|q(x,t) - \sum_{p=1}^{N} (-2\eta_p \operatorname{sgn}(\mu_p)\operatorname{sech}[2\eta_p(x-x_p^+-4\eta_p^2 t)])\right| = 0$$

with x_p^+ *as in (2.3b).*

Furthermore, besides the bound (5.3b), which depends only on t, we obtain from (4.2-3) and theorem 5.1 in the coordinate region $T = (3t)^{1/3} \geq 1$, $x \geq -\alpha$, $\alpha = \mu + \nu T$, the x and t dependent bound

$$(5.6a) \qquad |q(x,t) - q_d(x,t)| \leq aT^{-1}\exp\left[-\frac{2}{3}\,\theta\left(\frac{x+\alpha}{T}\right)^{3/2}\right] + bT^{-3}\left(1 + \frac{x+\alpha}{T}\right)^{-3/2}$$

$$(5.6b) \qquad a = \tilde{\gamma}\gamma^{-1}\gamma_\theta, \qquad b = \tilde{\gamma}\gamma^{-1}\rho$$

with $\theta \in (0,1)$, γ_θ, ρ the constants introduced in section 4.

It is a direct consequence of the last remark of Chapter 4, section 6, that the estimates (5.2-3-6) remain valid if on the left one replaces $|q(x,t) - q_d(x,t)|$ by

$$(5.7) \qquad \left| \int_x^\infty \left(q^2(s,t) - q_d^2(s,t) \right) ds \right|.$$

In particular, this yields

$$(5.8) \qquad \int_{-\alpha}^\infty q^2(x,t)dx = \int_{-\alpha}^\infty q_d^2(x,t)dx + O(t^{-1/3}) \qquad \text{as } t \to \infty.$$

Note that by (5.6)

$$(5.9) \qquad \int_{-\alpha}^\infty |q(x,t) - q_d(x,t)|^2 dx = O(t^{-1/3}) \qquad \text{as } t \to \infty.$$

This gives us another way to derive (5.8) (see Chapter 2, pp. 73, 74).

In the special case that the breather bound states satisfy (2.7) we find from (2.10b-12) and (5.8)

$$(5.10a) \qquad \int_{-\alpha}^\infty q^2(x,t)dx = 4 \sum_{p=1}^N \operatorname{Im} \zeta_p + O(t^{-1/3}) \qquad \text{as } t \to \infty.$$

Using the formula (see [2])

$$(5.11) \qquad \int_{-\infty}^\infty q^2(x,t)dx = \frac{2}{\pi} \int_0^\infty \log\left(1 + |b_r(\zeta)|^2\right) d\zeta + 4 \sum_{p=1}^N \operatorname{Im} \zeta_p$$

we obtain for the complementary integral

$$(5.10b) \qquad \int_{-\infty}^{-\alpha} q^2(x,t)dx = \frac{2}{\pi} \int_0^\infty \log\left(1 + |b_r(\zeta)|^2\right) d\zeta + O(t^{-1/3}) \text{ as } t \to \infty.$$

It is interesting to compare (5.10) with the formulae Chapter 2, (7.20-25) obtained in the KdV case.

Finally, let us remark that, as in Chapter 2, we can apply theorem 5.1 to obtain estimates in subregions of (1.4), e.g. $x \geq vt > 0$, with $v > 0$ arbitrarily fixed.

Specifically, if we make the additional assumption

(5.12) There is an integer $\tilde{n} \geq 2$ such that $b_r \in C^{\tilde{n}}(\mathbb{R})$ and all derivatives $b_r^{(j)}(\zeta)$, $j = 0,1,\ldots,\tilde{n}$ satisfy

$$b_r^{(j)}(\zeta) = \mathcal{O}(|\zeta|^{-2}), \qquad \zeta \to \pm\infty$$

then, reasoning as in Chapter 2, p. 75, we obtain

$$(5.13) \qquad \sup_{x \geq vt} |q(x,t) - q_d(x,t)| = \mathcal{O}(t^{-\tilde{n}}) \qquad \text{as } t \to \infty.$$

Likewise, one can make the extra assumption

(5.14) There is an $\varepsilon_0 > 0$ such that $q_0(x) = \mathcal{O}(\exp(-2\varepsilon_0 x))$ as $x \to +\infty$

and combine the remark at the end of Chapter 4, section 2, with the reasoning in Chapter 2, pp. 75, 76. Choosing $\varepsilon_1 > 0$ to be strictly less than ε_0, $\frac{1}{2}\sqrt{v}$ and $\operatorname{Im} \zeta_j$, $j = 1,2,\ldots,N$ one then arrives at

$$(5.15) \qquad \sup_{x \geq vt} |q(x,t) - q_d(x,t)| = \mathcal{O}(\exp(-\alpha_1 t)) \qquad \text{as } t \to \infty$$

with $\alpha_1 = 2\varepsilon_1(v - 4\varepsilon_1^2) > 0$.

References

[1] M.J. Ablowitz, D.J. Kaup, A.C. Newell and H. Segur, The inverse scattering transform - Fourier analysis for nonlinear problems, Stud. Appl. Math. 53 (1974), 249-315.

[2] M.J. Ablowitz and H. Segur, Solitons and the Inverse Scattering Transform, Philadelphia, SIAM 1981.

[3] Y. Kametaka, Korteweg-de Vries equation IV. Simplest generalization, Proc. Japan Acad., 45 (1969), 661-665.

[4] P.D. Lax, Integrals of nonlinear equations of evolution and solitary waves, Comm. Pure Appl. Math. 21 (1968), 467-490.

[5] G.L. Lamb Jr., Elements of Soliton Theory, Wiley-Interscience, 1980.

[6] M. Ohmiya, On the generalized soliton solutions of the modified Korteweg-de Vries equation, Osaka J. Math. 11 (1974), 61-71.

[7] P. Schuur, Asymptotic estimates of solutions of the Korteweg-de Vries equation on right half lines slowly moving to the left, preprint 330, Mathematical Institute Utrecht (1984).

[8] P. Schuur, On the approximation of a real potential in the Zakharov-Shabat system by its reflectionless part, preprint 341, Mathematical Institute Utrecht (1984).

[9] H. Segur and M.J. Ablowitz, Asymptotic solutions of nonlinear evolution equations and a Painlevé transcendent, Proc. Joint US-USSR Symposium on Soliton Theory, Kiev 1979, V.E. Zakharov and S.V. Manakov eds., North-Holland, Amsterdam, 165-184.

[10] S. Tanaka, Non-linear Schrödinger equation and modified Korteweg-de Vries equation; construction of solutions in terms of scattering data, Publ. R.I.M.S. Kyoto Univ. 10 (1975), 329-357.

[11] M. Wadati and K. Ohkuma, Multiple-pole solutions of the modified Korteweg-de Vries equation, J. Phys. Soc. Japan, 51 (6) (1982), 2029-2035.

CHAPTER SIX

MULTISOLITON PHASE SHIFTS FOR THE MODIFIED KORTEWEG-DE VRIES EQUATION IN THE CASE OF A NONZERO REFLECTION COEFFICIENT

We study multisoliton solutions of the modified Korteweg-de Vries equation in the case of a nonzero reflection coefficient. Confining ourselves to the case that all bound states are purely imaginary, we derive an explicit phase shift formula that clearly displays the nature of the interaction of each soliton with the other ones and with the dispersive wavetrain. In particular, this formula shows that each soliton experiences, in addition to the ordinary N-soliton phase shift, an extra phase shift to the right caused by the interaction with the dispersive wavetrain.

1. Introduction.

We consider the modified Korteweg-de Vries (mKdV) equation $q_t + 6q^2q_x + q_{xxx} = 0$ with arbitrary real initial conditions $q(x,0) = q_0(x)$, such that q_0 is a bona fide potential in the Zakharov-Shabat scattering problem, sufficiently smooth and rapidly decaying for $|x| \to \infty$ for the whole of the Zakharov-Shabat inverse scattering method

[1] to work and to guarantee certain regularity and decay properties of the scattering data, to be stated further on.

The long-time behaviour of the solution $q(x,t)$ of the mKdV problem has been treated by several authors. For details we refer to [9], i.e. Chapter 5, where some of the idiosyncrasies are discussed.

In the asymptotic part of this chapter (sections 3, 4 and 5) we confine ourselves to the case that all bound states are purely imaginary. Then it is found that as $t \to +\infty$ the solution decomposes into N sech-shaped solitons moving to the right and a dispersive wavetrain moving to the left. As $t \to -\infty$ the arrangement is reversed. One can now ask for the phase shifts of the solitons as they interact both with the other solitons and with the dispersive wavetrain.

In this chapter, starting from our asymptotic analysis of the solution given in Chapter 5, we derive a phase shift formula that closely resembles that found by Ablowitz and Kodama [2] for the KdV case. We next show that - as in the KdV case - a simple substitution produces a more transparent formula. From the latter formula it is evident, that each soliton experiences, in addition to the ordinary N-soliton phase shift, an extra phase shift to the right, the so-called continuous phase shift, caused by the interaction with the dispersive wavetrain. Thus, the presence of reflection causes an advancement in the soliton motion. Note that in our KdV analysis [8], i.e. Chapter 3, we found the opposite. There the interaction with the dispersive wavetrain causes a delay in the soliton motion, since the continuous phase shifts are to the left.

The composition of this chapter closely resembles that of Chapter 3. In section 2 we briefly discuss the left and right scattering data associated with $q_0(x)$ and show how, in the general case that the bound states are distributed symmetrically with respect to the imaginary axis, the left scattering data can be expressed in terms of the right scattering data in a convenient way. In the rest of the chapter we assume that all bound states are purely imaginary. In section 3 we quote a result from Chapter 5, describing the asymptotic behaviour of $q(x,t)$ as $t \to +\infty$. By the same symmetry argument as in Chapter 3, we derive from this result the asymptotic behaviour of $q(x,t)$ as $t \to -\infty$. Next, in section 4, the two asymptotic results are combined to give a phase shift formula of Ablowitz-

Kodama type. The representation of the left normalization coefficients in terms of the right scattering data, which was obtained in section 2, then enables us to write the phase shift formula in a more transparent form. To illustrate our results, we calculate in section 5 the continuous phase shifts arising from a sech initial function.

2. Left and right scattering data and their relationship.

For Im $\zeta \geq 0$ we introduce the Jost functions $\psi_r(x,\zeta)$ and $\psi_\ell(x,\zeta)$, two special solutions of the Zakharov-Shabat system

$$(2.1) \qquad \begin{pmatrix}\psi_1\\ \psi_2\end{pmatrix}' = \begin{pmatrix}-i\zeta & q_0\\ -q_0 & i\zeta\end{pmatrix}\begin{pmatrix}\psi_1\\ \psi_2\end{pmatrix}, \qquad ' = \frac{d}{dx}, \quad -\infty < x < +\infty,$$

uniquely determined by

$$(2.2a) \qquad \psi_r(x,\zeta) = e^{-i\zeta x}R(x,\zeta), \qquad \lim_{x\to-\infty} R(x,\zeta) = \begin{pmatrix}1\\ 0\end{pmatrix}$$

$$(2.2b) \qquad \psi_\ell(x,\zeta) = e^{i\zeta x}L(x,\zeta), \qquad \lim_{x\to+\infty} L(x,\zeta) = \begin{pmatrix}0\\ 1\end{pmatrix}.$$

Here R and L are vector functions with two components. We set

$$(2.3a) \qquad r_-(\zeta) = 1 + \int_{-\infty}^{\infty} q_0(x)R_2(x,\zeta)dx, \qquad \text{Im } \zeta \geq 0$$

$$(2.3b) \qquad r_+(\zeta) = -\int_{-\infty}^{\infty} q_0(x)e^{-2i\zeta x}R_1(x,\zeta)dx, \qquad \zeta \in \mathbb{R}$$

$$(2.3c) \qquad \ell_+(\zeta) = 1 + \int_{-\infty}^{\infty} q_0(x)L_1(x,\zeta)dx, \qquad \text{Im } \zeta \geq 0$$

$$(2.3d) \qquad \ell_-(\zeta) = -\int_{-\infty}^{\infty} q_0(x)e^{2i\zeta x}L_2(x,\zeta)dx, \qquad \zeta \in \mathbb{R}.$$

Note that $r_-(\zeta) = \ell_+(\zeta)$ and $r_+(\zeta) = \ell_-(-\zeta)$. It is shown in [5] that $r_-(\zeta)$ is analytic on Im $\zeta > 0$ and continuous on Im $\zeta \geq 0$.

Following [5] and Chapter 5, we shall assume throughout that r_- has no zeros on the real axis and that all zeros of r_- in $\mathbb{C}_+$ are simple. As a result the function $r_-(\zeta)$ has at most finitely many zeros $\zeta_1,\zeta_2,\ldots,\zeta_N$,

Im $\zeta_m > 0$, distributed symmetrically with respect to the imaginary axis. We shall call them the bound states associated with q_0. From [5], p. 156 we know that there are nonzero constants α_m such that

$$\psi_r(x,\zeta_m) = \alpha_m \psi_\ell(x,\zeta_m). \tag{2.4}$$

Furthermore, by [5], (5.3.2) one has

$$\left.\frac{dr_-}{d\zeta}\right|_{\zeta=\zeta_m} = 2(i\alpha_m)^{-1} \int_{-\infty}^{\infty} \psi_{r_1}(x,\zeta_m)\psi_{r_2}(x,\zeta_m)dx. \tag{2.5}$$

Observe that the integral does not vanish since the ζ_m are simple zeros of r_-. This enables us to introduce the right normalization coefficients

$$C_m^r = \tfrac{1}{2}i\left[\int_{-\infty}^{\infty} \psi_{\ell_1}(x,\zeta_m)\psi_{\ell_2}(x,\zeta_m)dx\right]^{-1}, \tag{2.6a}$$

as well as the left normalization coefficients

$$C_m^\ell = \tfrac{1}{2}i\left[\int_{-\infty}^{\infty} \psi_{r_1}(x,\zeta_m)\psi_{r_2}(x,\zeta_m)dx\right]^{-1}. \tag{2.6b}$$

Next, we introduce the following functions of $\zeta \in \mathbb{R}$

(2.7a) $a_r = r_-^{-1}$, the right transmission coefficient,

(2.7b) $a_\ell = \ell_+^{-1}$, the left transmission coefficient,

(2.7c) $b_r = r_+ r_-^{-1}$, the right reflection coefficient,

(2.7d) $b_\ell = \ell_- \ell_+^{-1}$, the left reflection coefficient.

We shall call the aggregate of quantities $\{a_r(\zeta), b_r(\zeta), \zeta_m, C_m^r\}$ the right scattering data associated with the potential q_0. Similarly we refer to $\{a_\ell(\zeta), b_\ell(\zeta), \zeta_m, C_m^\ell\}$ as the left scattering data associated with q_0.
Note that the right scattering data were already introduced in Chapter 4, where some of their properties were discussed.
We claim that a_ℓ, b_ℓ and C_m^ℓ can be expressed in terms of the right scattering data in the following way

$$a_\ell(\zeta) = a_r(\zeta) \qquad b_\ell(\zeta) = \frac{a_r(\zeta)}{a_r(-\zeta)}\, b_r(-\zeta), \tag{2.8a}$$

(2.8b) $$c_m^{\ell} = [c_m^r]^{-1}(\zeta_m - \zeta_m^*)^2\left\{\exp\left\{\frac{2\zeta_m}{\pi i}\int_0^{\infty}\frac{\log(1+|b_r(\zeta)|^2)}{\zeta^2-\zeta_m^2}\,d\zeta\right\}\right\}$$

$$\times \prod_{\substack{p=1\\ p\neq m}}^{N}\left(\frac{\zeta_m-\zeta_p^*}{\zeta_m-\zeta_p}\right)^2 .$$

Clearly, only formula (2.8b) deserves a proof.
To provide it, note first that by (2.4-6)

(2.9) $$c_m^r = \alpha_m^2 c_m^{\ell}.$$

Next, it follows from (2.5-6b) that

(2.10) $$\left.\frac{dr_-}{d\zeta}\right|_{\zeta=\zeta_m} = [\alpha_m c_m^{\ell}]^{-1}.$$

Eliminating α_m from (2.9) and (2.10) we find

(2.11) $$c_m^r c_m^{\ell}\left[\left.\frac{dr_-}{d\zeta}\right|_{\zeta=\zeta_m}\right]^2 = 1.$$

Lastly, from [3], p. 57 we obtain the representation

(2.12) $$r_-(\zeta) = \left\{\exp\left\{\frac{1}{2\pi i}\int_{-\infty}^{\infty}\frac{\log(1+|b_r(\omega)|^2)}{\zeta-\omega}\,d\omega\right\}\right\}\prod_{p=1}^{N}\frac{\zeta-\zeta_p}{\zeta-\zeta_p^*}, \quad \mathrm{Im}\,\zeta > 0.$$

Differentiating and using the symmetry relation $b_r^*(\zeta) = b_r(-\zeta)$, we find

(2.13) $$\left.\frac{dr_-}{d\zeta}\right|_{\zeta=\zeta_m} = (\zeta_m-\zeta_m^*)^{-1}\left\{\exp\left\{\frac{-\zeta_m}{\pi i}\int_0^{\infty}\frac{\log(1+|b_r(\zeta)|^2)}{\zeta^2-\zeta_m^2}\,d\zeta\right\}\right\}\prod_{\substack{p=1\\ p\neq m}}^{N}\frac{\zeta_m-\zeta_p}{\zeta_m-\zeta_p^*} .$$

Together, (2.11) and (2.13) yield the desired formula (2.8b).

3. Forward and backward asymptotics.

Once the right scattering data of $q_0(x)$ are known, the solution $q(x,t)$ of the forward mKdV problem

(3.1) $$\begin{cases} q_t + 6q^2q_x + q_{xxx} = 0, & -\infty < x < +\infty, \quad t > 0 \\ q(x,0) = q_0(x) \end{cases}$$

can in principle be computed by the inverse scattering method [1]. An asymptotic analysis of the solution was presented in Chapter 5. It was found there that the asymptotic structure of the reflectionless part of $q(x,t)$ is rather complicated when the location of the bound states ζ_m is not suitably restricted.

To avoid any unpleasantness of this kind we shall assume from now on that all the bound states are purely imaginary. Let us denote them by

(3.2) $$\zeta_m = i\eta_m, \quad \text{with } 0 < \eta_N < \dots < \eta_2 < \eta_1.$$

Since $\psi_r(x,i\eta_m)$ and $\psi_\ell(x,i\eta_m)$ are real we then have that the normalization coefficients are purely imaginary as well, say

(3.3) $$C_m^r = i\mu_m^r, \quad C_m^\ell = i\mu_m^\ell, \quad \text{with } \mu_m^r, \mu_m^\ell \in \mathbb{R}\setminus\{0\}.$$

The asymptotic behaviour of the solution $q(x,t)$ of (3.1) is now enlightened by the following lemma, which was established in Chapter 5, section 5.

Lemma 3.1. *Assume that*

(3.4) $b_r(\zeta)$ *is of class* $C^2(\mathbb{R})$ *and the derivatives* $b_r^{(j)}(\zeta)$, $j = 0,1,2$ *are bounded on* $\mathbb{R}$.

Then one has

(3.5) $$\lim_{t\to\infty} \sup_{x\geq -t^{1/3}} \left| q(x,t) - \sum_{m=1}^{N} \left\{-2\eta_m \operatorname{sgn}(\mu_m^r)\operatorname{sech}[2\eta_m(x-x_m^+-4\eta_m^2 t)]\right\}\right| = 0,$$

where

(3.6) $$x_m^+ = \frac{1}{2\eta_m}\log\left\{\frac{|\mu_m^r|}{2\eta_m}\prod_{p=1}^{m-1}\left(\frac{\eta_p-\eta_m}{\eta_p+\eta_m}\right)^2\right\}.$$

Next, let us consider the backward mKdV problem, starting from the same initial function $q_0(x)$, i.e.

(3.7) $$\begin{cases} q_t + 6q^2 q_x + q_{xxx} = 0, & -\infty < x < +\infty, \quad t < 0 \\ q(x,0) = q_0(x). \end{cases}$$

Plainly, if $q(x,t)$ satisfies (3.7), then $w(x,t) = q(-x,-t)$ satisfies

$$(3.8) \qquad \begin{cases} w_t + 6w^2 w_x + w_{xxx} = 0, & -\infty < x < +\infty, \quad t > 0 \\ w(x,0) = q_0(-x), \end{cases}$$

so that $w(x,t)$ satisfies the forward mKdV problem with initial function $q_0(-x)$. To solve (3.7) it is therefore sufficient to determine the right scattering data associated with $q_0(-x)$ and apply the inverse scattering method to (3.8). However, an easy calculation reveals that the right scattering data associated with $q_0(-x)$ are identical with the left scattering data associated with $q_0(x)$. The latter were examined in the previous section. Thus, to find the asymptotic behaviour of the solution $q(x,t)$ of (3.7) for $t \to -\infty$ we merely apply lemma 3.1 to problem (3.8) and perform the transcription $q(x,t) = w(-x,-t)$. This yields

Lemma 3.2. *Assume that*

(3.9) $b_\ell(\zeta)$ *is of class* $C^2(\mathbb{R})$ *and the derivatives* $b_\ell^{(j)}(\zeta)$, $j = 0,1,2$ *are bounded on* $\mathbb{R}$.

Then one has

$$(3.10) \qquad \lim_{t\to-\infty} \sup_{x \le |t|^{1/3}} \left| q(x,t) - \sum_{m=1}^{N} \left\{-2\eta_m \operatorname{sgn}(\mu_m^\ell) \operatorname{sech}[2\eta_m(x - x_m^- - 4\eta_m^2 t)]\right\}\right| = 0,$$

where

$$(3.11) \qquad x_m^- = -\frac{1}{2\eta_m} \log\left\{\frac{|\mu_m^\ell|}{2\eta_m} \prod_{p=1}^{m-1} \left(\frac{\eta_p - \eta_m}{\eta_p + \eta_m}\right)^2\right\}.$$

Since α_m is real it follows from (2.9) that $\operatorname{sgn}(\mu_m^r) = \operatorname{sgn}(\mu_m^\ell)$. Therefore, in (3.5) and (3.10) the same solitons emerge, apart from a phase shift.

4. An explicit phase shift formula.

Let us assume that b_r and b_ℓ satisfy the conditions (3.4) and (3.9). Then the convergence results (3.5) and (3.10) display clearly how the solution $q(x,t)$ of the mKdV equation evolving from $q(x,0) = q_0(x)$ splits up into N solitons as $t \to \pm\infty$. In particular, we find for the m-th soliton the following phase shift

$$(4.1)\qquad S_m = x_m^+ - x_m^- = \frac{1}{2\eta_m} \log\left\{\frac{|\mu_m^r \mu_m^\ell|}{4\eta_m^2} \prod_{p=1}^{m-1} \left(\frac{\eta_p - \eta_m}{\eta_p + \eta_m}\right)^4\right\}.$$

Note how closely this resembles the phase shift formula derived by Ablowitz and Kodama [2] for the KdV case (see Chapter 3, (4.1)).

Similarly, the formulae (3.11) and (4.1) become more transparent if one inserts the representation (2.8b), taking into account (3.2-3). In summary, this leads to

$$(4.2a)\qquad x_m^+ = \frac{1}{2\eta_m} \log\left(\frac{|\mu_m^r|}{2\eta_m}\right) + \frac{1}{\eta_m} \sum_{p=1}^{m-1} \log\left(\frac{\eta_p - \eta_m}{\eta_p + \eta_m}\right)$$

$$(4.2b)\qquad x_m^- = \frac{1}{2\eta_m} \log\left(\frac{|\mu_m^r|}{2\eta_m}\right) + \frac{1}{\eta_m} \sum_{p=m+1}^{N} \log\left(\frac{\eta_m - \eta_p}{\eta_m + \eta_p}\right) - \frac{1}{\pi} \int_0^\infty \frac{\log(1+|b_r(\zeta)|^2)}{\zeta^2 + \eta_m^2} d\zeta$$

$$(4.3a)\qquad S_m = S_m^d + S_m^c$$

$$(4.3b)\qquad S_m^d = \frac{1}{\eta_m} \sum_{p=1}^{m-1} \log\left(\frac{\eta_p - \eta_m}{\eta_p + \eta_m}\right) - \frac{1}{\eta_m} \sum_{p=m+1}^{N} \log\left(\frac{\eta_m - \eta_p}{\eta_m + \eta_p}\right)$$

$$(4.3c)\qquad S_m^c = \frac{1}{\pi} \int_0^\infty \frac{\log(1 + |b_r(\zeta)|^2)}{\zeta^2 + \eta_m^2} d\zeta .$$

In S_m^d we recognize the pure N-soliton phase shift (caused by pairwise interaction of the m-th soliton with the other ones). The quantity S_m^c (which is <u>positive</u> for nonzero b_r) can be seen as the shift caused by the interaction of the m-th soliton with the dispersive wavetrain. For nonzero b_r we obviously have

$$(4.4)\qquad S_N^c > \ldots > S_2^c > S_1^c > 0.$$

Thus, surprisingly enough, the interaction with the dispersive wavetrain <u>advances</u> the solitons in their motion and the effect is most heavily felt by the smallest one, corresponding to η_N. Recall that in our KdV analysis (Chapter 3) we found the opposite situation. There the interaction with the dispersive wavetrain causes a delay in the motion of the solitons.

Let us examine where this difference comes from. Apparently, formula

(4.3b) for the mKdV pure N-soliton phase shift S_m^d coincides with the corresponding KdV formula Chapter 3, (4.3b), provided we identify η_m with κ_m, $m = 1,2,\dots,N$. But after this identification it is clear that, remarkably enough, also the continuous phase shifts S_m^c are given by the same formula, namely

$$(4.5) \qquad S_m^c = \frac{1}{\pi} \int_0^\infty \frac{\log|a_r(k)|^2}{k^2 + \kappa_m^2}\, dk,$$

where $a_r(k)$ denotes the right transmission coefficient. Now, suppose b_r is not identically zero. Then, in the KdV case we have by Chapter 2, (2.12)

$$(4.6a) \qquad |a_r(k)|^2 = 1 - |b_r(k)|^2,$$

so that S_m^c is negative.

By contrast, in the mKdV case, Chapter 4, (2.24b) tells us

$$(4.6b) \qquad |a_r(\zeta)|^2 = 1 + |b_r(\zeta)|^2,$$

leading to a positive sign of S_m^c. Thus the difference in sign of S_m^c stems from the difference in sign of $|b_r(k)|^2$ in the formulae (4.6).

Using the formula (see [3])

$$(4.7) \qquad \int_{-\infty}^\infty q_0^2(x)dx = \frac{2}{\pi} \int_0^\infty \log(1 + |b_r(\zeta)|^2)d\zeta + 4 \sum_{p=1}^N \operatorname{Im} \zeta_p$$

we obtain for the continuous phase shift S_m^c the following estimate in terms of the initial function $q_0(x)$ and the bound states $\zeta_p = i\eta_p$:

$$(4.8) \qquad 0 \le S_m^c \le \frac{1}{2\eta_m^2}\left(\int_{-\infty}^\infty q_0^2(x)dx - 4 \sum_{p=1}^N \eta_p\right).$$

5. An example: the continuous phase shifts arising from a sech initial function.

To make the previous discussion less abstract let us compute the continuous phase shifts arising from the initial function

(5.1) $$q_0(x) = \alpha \operatorname{sech} x, \quad \alpha \in \mathbb{R}\setminus\{k+\tfrac{1}{2};\ k \in \mathbb{Z}\}.$$

It is a remarkable fact that for the potential (5.1) one can solve the scattering problem (2.1) in closed form. In fact by an obvious modification of the calculation performed in [7], section 3, we obtain for $x \in \mathbb{R}$, $\operatorname{Im} \zeta \geq 0$

(5.2a) $$R_1(x,\zeta) = F(\alpha,-\alpha;\ \tfrac{1}{2}-i\zeta;\ z)$$

(5.2b) $$R_2(x,\zeta) = \alpha^{-1} z^{\frac{1}{2}}(1-z)^{\frac{1}{2}} \frac{d}{dz} F(\alpha,-\alpha;\ \tfrac{1}{2}-i\zeta;\ z)$$

with

(5.3) $$z = \tfrac{1}{2}(1 + \tanh x).$$

Here F denotes the hypergeometric function in the notation of [4], p. 556. The same argument as in [7] now gives us

(5.4a) $$r_-(\zeta) = \frac{\Gamma^2(\frac{1}{2}-i\zeta)}{\Gamma(\frac{1}{2}+\alpha-i\zeta)\Gamma(\frac{1}{2}-\alpha-i\zeta)}, \qquad \operatorname{Im} \zeta \geq 0$$

(5.4b) $$r_+(\zeta) = -\frac{\sin \pi\alpha}{\cosh \pi\zeta}, \qquad \zeta \in \mathbb{R}.$$

Note that the assumptions about r_- made in sections 2 and 3 are fulfilled since all zeros of $r_-(\zeta)$ in $\operatorname{Im} \zeta \geq 0$ are simple and lie on the positive imaginary axis.

Specifically, one has the following situation.

If $|\alpha| < \frac{1}{2}$, then $r_-(\zeta)$ has no zeros at all. Note that in this case

(5.5) $$\int_{-\infty}^{\infty} |q_0(x)|dx < \frac{\pi}{2},$$

which can be seen as a slight relaxation of the sufficient condition [1]

(5.6) $$\int_{-\infty}^{\infty} |q_0(x)|dx < 0.904$$

for the absence of solitons.

For $|\alpha| > \frac{1}{2}$, let $N \geq 1$ denote the unique integer such that $N - \frac{1}{2} < |\alpha| < N + \frac{1}{2}$. Then $r_-(\zeta)$ has precisely N zeros given by

(5.7) $$\zeta_p = i(\tfrac{1}{2} + |\alpha| - p), \quad p = 1,2,\ldots,N.$$

Clearly, $q_0(x)$ is reflectionless (i.e. $b_r \equiv 0$) if and only if $\alpha \in \mathbb{Z}$.

Moreover, if $\alpha \in \mathbb{Z}\setminus\{0\}$, then $q_0(x)$ is reflectionless with $N = |\alpha|$ bound states.

As is easily seen, q_0 belongs to the Schwartz space. Hence, the same holds for b_r, which equals b_ℓ, since q_0 is even. We conclude that (3.4) and (3.9) hold, so that (3.5) and (3.10) are valid.

Now, let us compute the continuous phase shifts S_m^c.
By (5.4a) one has

$$(5.8) \qquad r_-(i\nu) = \frac{\Gamma^2(\frac{1}{2}+\nu)}{\Gamma(\frac{1}{2}+\alpha+\nu)\Gamma(\frac{1}{2}-\alpha+\nu)}, \qquad \nu > 0.$$

On the other hand, by (2.12), (3.2)

$$(5.9) \qquad r_-(i\nu) = \left\{\exp\left\{-\frac{\nu}{\pi}\int_0^\infty \frac{\log(1+|b_r(\zeta)|^2)}{\zeta^2+\nu^2}\,d\zeta\right\}\right\}\prod_{p=1}^{N}\frac{\nu-\eta_p}{\nu+\eta_p}, \qquad \nu > 0.$$

Equating both expressions and making repeated use of the recurrence formula $\Gamma(z+1) = z\Gamma(z)$, we obtain

$$(5.10) \qquad \frac{1}{\pi}\int_0^\infty \frac{\log(1+|b_r(\zeta)|^2)}{\zeta^2+\nu^2}\,d\zeta = \frac{1}{\nu}\log\left\{\frac{\Gamma(\frac{1}{2}+|\alpha|+\nu-N)\Gamma(\frac{1}{2}-|\alpha|+\nu+N)}{\Gamma^2(\frac{1}{2}+\nu)}\right\}$$
$$= \frac{1}{\nu}\log\left\{\frac{B(\frac{1}{2}+|\alpha|+\nu-N,\frac{1}{2}-|\alpha|+\nu+N)}{B(\frac{1}{2}+\nu,\ \frac{1}{2}+\nu)}\right\},$$

where B refers to the beta function ([4], p. 258). Combining (5.7) and (5.10), we arrive at the following expression for the continuous phase shifts S_m^c

$$(5.11) \qquad S_m^c = \frac{1}{\frac{1}{2}+|\alpha|-m}\log\left\{\frac{B(1+2|\alpha|-m-N,\ 1-m+N)}{B(1+|\alpha|-m,\ 1+|\alpha|-m)}\right\}.$$

To estimate the magnitude of S_m^c we may benefit from (4.8), which yields

$$(5.12) \qquad 0 \le S_m^c \le \left(\frac{N-|\alpha|}{\frac{1}{2}+|\alpha|-m}\right)^2.$$

The lower bound in (5.12) can be improved by means of the inequality

$$(5.13) \qquad \frac{B(c-2b,\ c)}{B(c-b,c-b)} \ge 1+\frac{b^2}{c}, \qquad c > 0, \quad b < \tfrac{1}{2}c,$$

due to Gurland [6]. Together with the simple estimate $\log(1+x) \ge \frac{7}{8}x$ for $0 \le x \le \frac{1}{4}$, this tells us

$$(5.14) \qquad S_m^c \geq \frac{7}{8} \frac{(N - |\alpha|)^2}{(1 - m + N)(\frac{1}{2} + |\alpha| - m)} .$$

References

[1] M.J. Ablowitz, D.J. Kaup, A.C. Newell and H. Segur, The inverse scattering transform. Fourier analysis for nonlinear problems, Stud. Appl. Math. 53 (1974), 249-315.

[2] M.J. Ablowitz and Y. Kodama, Note on asymptotic solutions of the Korteweg-de Vries equation with solitons, Stud. Appl. Math. 66 (1982), No. 2, 159-170.

[3] M.J. Ablowitz and H. Segur, Solitons and the Inverse Scattering Transform, Philadelphia, SIAM, 1981.

[4] M. Abramowitz and I.A. Stegun, Handbook of Mathematical Functions, National Bureau of Standards Applied Mathematics Series, No. 55, U.S. Department of Commerce, 1964.

[5] W. Eckhaus and A. van Harten, The Inverse Scattering Transformation and the Theory of Solitons, North-Holland Mathematics Studies 50, 1981.

[6] J. Gurland, An inequality satisfied by the gamma function, Skand. Aktuarietidskr. 39 (1956), 171-172.

[7] J.W. Miles, An envelope soliton problem, SIAM J. Appl. Math. 41 (1981), No. 2, 227-230.

[8] P. Schuur, Multisoliton phase shifts in the case of a nonzero reflection coefficient, Phys. Lett. 102A (1984), No. 9, 387-392.

[9] P. Schuur, Decomposition and estimates of solutions of the modified Korteweg-de Vries equation on right half lines slowly moving leftward, preprint 342, Mathematical Institute Utrecht (1984).

CHAPTER SEVEN

ASYMPTOTIC ESTIMATES OF SOLUTIONS OF THE SINE-GORDON EQUATION ON RIGHT HALF LINES ALMOST LINEARLY MOVING LEFTWARD

We consider the sine-Gordon equation $q_t = \frac{1}{2}\sin[2 \int_{-\infty}^{x} q(x',t)dx']$ with arbitrary real initial conditions $q(x,0) = q_0(x)$, sufficiently smooth and rapidly decaying as $|x| \to \infty$, such that q_0 is a bona fide potential in the Zakharov-Shabat scattering problem. Using the method of the inverse scattering transformation we analyse the behaviour of the solution $q(x,t)$ in coordinate regions of the form $t > 0$, $x \geq -\mu - \nu t^{\delta}$, where μ, ν and δ are nonnegative constants with $\delta < 1$.
We derive explicit x and t dependent bounds for the non-reflectionless part of $q(x,t)$. Owing to the rather explicit structure of the reflectionless part it is then a small step to obtain estimates of $q(x,t)$ as well as some interesting energy formulae.

1. Introduction.

We study the sine-Gordon problem

(1.1a) $$q_t = \frac{1}{2}\sin\left[2 \int_{-\infty}^{x} q(x',t)dx'\right], \qquad -\infty < x < +\infty, \qquad t > 0$$

(1.1b) $$q(x,0) = q_0(x),$$

where the initial function $q_0(x)$ is an arbitrary real function on $\mathbb{R}$, such that

(1.2a) $q_0(x)$ satisfies the hypotheses (2.2-13) in [9] (i.e. Chapter 4) and is therefore a bona fide potential in the Zakharov-Shabat scattering problem.

(1.2b) There is an integer k_0 such that

$$\int_{-\infty}^{\infty} q_0(x)dx = k_0\pi.$$

(1.2c) $q_0(x)$ is sufficiently smooth and (along with a number of its derivatives) decays sufficiently rapidly for $|x| \to \infty$:

(i) for the whole of the Zakharov-Shabat inverse scattering method [1], [2] to work,

(ii) to guarantee certain regularity and decay properties of the right reflection coefficient to be stated further on.

Uniqueness of solutions of (1.1) can be proven within the class of functions $q(x,t)$ vanishing sufficiently rapidly for $|x| \to \infty$ and satisfying

$$\int_{-\infty}^{\infty} q(x,t)dx = k_0\pi. \qquad (1.3)$$

Suitably adapting the procedure outlined in [11] for the mKdV problem one can establish by an inverse scattering analysis that condition (1.2) guarantees the existence of a real function $q(x,t)$, continuous on $\mathbb{R}\times[0,\infty)$, such that

(1.4a) For each $t \geq 0$ the function $q(x,t)$ satisfies the hypotheses (2.2-13) in Chapter 4.

(1.4b) $q(x,t)$ has the property (1.3).

(1.4c) $q(x,t)$ satisfies (1.1) in the classical sense.

(1.4d) $q(x,t)$ falls in the class of functions for which uniqueness of solutions of (1.1) can be proven.

Whenever, in this chapter, we speak of "the solution" of (1.1) we shall refer to the solution obtained by inverse scattering. Let us add that, despite the multitude of sine-Gordon papers appeared so far, we know of no reference in which the above is spelled out in full detail.

Given the solution $q(x,t)$ of (1.1) we put

$$\text{(1.5)} \qquad \sigma(x,t) = -2 \int_{-\infty}^{x} q(x',t)dx', \qquad \sigma_0(x) = -2 \int_{-\infty}^{x} q_0(x')dx'.$$

Plainly, $\sigma(x,t)$ satisfies

$$\text{(1.6a)} \qquad \sigma_{xt} = \sin \sigma, \qquad -\infty < x < +\infty, \qquad t > 0$$

$$\text{(1.6b)} \qquad \sigma(x,0) = \sigma_0(x)$$

and for fixed $t \geq 0$ one has

$$\text{(1.7)} \qquad \lim_{x\to-\infty} \sigma(x,t) = 0, \qquad \lim_{x\to+\infty} \sigma(x,t) = -2k_0\pi.$$

In discussions based on the Zakharov-Shabat inverse scattering method the version (1.6) of the sine-Gordon problem is most frequently used (see [1], [2], [5], [7]), the only cases allowing an easy treatment being those in which σ has the boundary behaviour displayed in (1.7).

Let us recall that by the inverse scattering method the solution $q(x,t)$ of (1.1) is obtained in the following way.
First one computes the (right) scattering data $\{b_r(\zeta),\zeta_j,C_j^r\}$ associated with $q_0(x)$. For their definition and properties we refer to Chapter 4. Next one puts (see [1], [2])

$$\text{(1.8a)} \qquad C_j^r(t) = C_j^r\exp\{-it/(2\zeta_j)\}, \qquad j = 1,2,\ldots,N$$

$$\text{(1.3b)} \qquad b_r(\zeta,t) = b_r(\zeta)\exp\{-it/(2\zeta)\}, \qquad -\infty < \zeta < +\infty.$$

Then by the solvability of the inverse scattering problem, there exists for each $t > 0$ a real potential $q(x,t)$, satisfying the hypotheses (2.2-13) in Chapter 4 and having $\{b_r(\zeta,t),\zeta_j,C_j^r(t)\}$ as its scattering data. Note that by Chapter 4, (2.25)

$$\text{(1.9)} \qquad b_r(0,t) = -\tan\left[\int_{-\infty}^{\infty} q(x,t)dx\right],$$

so that property (1.3) follows from (1.1b-2b-8b-9) and a continuity argument. The function $q(x,t)$ is the unique solution of the sine-Gordon initial value problem (1.1).

Explicit solutions by the above procedure have only been obtained for $b_r \equiv 0$. The solution $q_d(x,t)$ of the sine-Gordon equation (1.1a) with scattering data $\{0,\zeta_j,C_j^r(t)\}$ is called the reflectionless part of $q(x,t)$.

By contrast with KdV and mKdV, the long-time behaviour of the solution $q(x,t)$ of the sine-Gordon problem (1.1) is not treated extensively in the literature.

In the absence of solitons, it is suggested in [3], pp. 90, 91, that there are three distinct asymptotic regions

I. $x \geq \tilde{O}(t)$ II. $|x| \leq \tilde{O}(t)$
III. $-x \geq \tilde{O}(t)$,

where $\tilde{O}$ denotes positive proportionality, within each of which the solution $q(x,t)$ of (1.1) has a different asymptotic expansion. Note that, in view of the negative group velocity associated with the linearized version of (1.1), one may expect that the solution will evolve into a dispersive wavetrain moving to the left.

If solitons are present, then, generically, as time goes on, $q_d(x,t)$ will desintegrate into breathers and sech-shaped solitons (see the discussion in section 2). But the sech-shaped solitons as well as the breather envelopes propagate to the left. Consequently, a decomposition of $q(x,t)$ into a dispersive and a soliton part will not easily be demonstrated in this case. The only thing beyond doubt is that for large t the solution $q(x,t)$ will be confined to the negative x-axis. This raises the question: How small is $q(x,t)$ in the regions I and II?

In this chapter we give a definite answer to this question by analysing the behaviour of $q(x,t)$ in coordinate regions of the form

$$(1.10) \qquad t > 0, \quad x \geq -\alpha, \quad \alpha = \mu + \nu t^{\delta}$$

where μ, ν and δ are nonnegative constants with $\delta < 1$. Here μ and ν are arbitrary but δ depends on properties of q_0.
More precisely, let q_0 and a number of its derivatives decay fast enough to guarantee that b_r has $n \geq 2$ derivatives decaying sufficiently rapidly (see (4.1)).

Choose

$$(1.11) \qquad 0 < \lambda \leqq n - \frac{3}{2}, \qquad 0 \leqq \delta \leqq 1 - \left(\frac{3 + 2\lambda}{2n}\right).$$

Then we prove that

$$(1.12) \qquad \sup_{x \geqq -\alpha} |q(x,t)| = O(t^{-\lambda}) \text{ as } t \to \infty.$$

Moreover, we derive the energy formulae

$$(1.13a) \qquad \int_{-\alpha}^{\infty} q^2(x,t)dx = O(t^{-\lambda}) \quad \text{as } t \to \infty$$

$$(1.13b) \qquad \int_{-\infty}^{-\alpha} q^2(x,t)dx = \frac{2}{\pi} \int_0^{\infty} \log(1 + |b_r(\zeta)|^2)d\zeta + 4 \sum_{p=1}^{N} \operatorname{Im} \zeta_p + O(t^{-\lambda})$$

as $t \to \infty$.

In particular, if q_0 is in the Schwartz class then (1.12-13) hold for all $\lambda > 0$ and all $0 \leqq \delta < 1$.

The chapter runs as follows.
In section 2 we isolate certain properties of the reflectionless part $q_d(x,t)$. In section 3 we recall a theorem established previously in Chapter 4, in which $q(x,t) - q_d(x,t)$ is estimated in terms of $\Omega_c(x+y;t)$. Section 4 is devoted to the construction of some simple explicit bounds of $\Omega_c(x+y;t)$. Then in section 5 these bounds and the theorem are combined to yield estimates of $q(x,t) - q_d(x,t)$. Since $q_d(x,t)$ was already explored in section 2, we have reached our goal and found estimates of $q(x,t)$.

2. The asymptotic structure of the reflectionless part.

From Chapter 4, (5.28) we obtain for the reflectionless part of $q(x,t)$ the explicit expression

$$(2.1) \qquad q_d(x,t) = 2 \operatorname{Im} \frac{\partial}{\partial x} \log \det A,$$

where $A = (\alpha_{pj})$ denotes the N×N matrix with elements

$$(2.2) \qquad \alpha_{pj} = [c_j^r(t)]^{-1} e^{-2i\zeta_j x} \delta_{pj} + i(\zeta_p + \zeta_j)^{-1}.$$

If $M = 0$ in Chapter 4, (2.18), then the asymptotic structure of $q_d(x,t)$ is relatively simple. In that case all the bound states and normalization coefficients are purely imaginary, say $\zeta_j = i\eta_j$, $0 < \eta_N < \ldots < \eta_2 < \eta_1$ and $C_j^r = i\mu_j$, $\mu_j \in \mathbb{R}\backslash\{0\}$. An obvious adaptation of the reasoning given in [8] shows that as t approaches infinity then the reflectionless part of q(x,t) decomposes into N solitons uniformly with respect to x on $\mathbb{R}$. Specifically

$$\text{(2.3a)} \qquad \lim_{t\to\infty} \sup_{x\in\mathbb{R}} \left| q_d(x,t) - \sum_{p=1}^{N} \left(-2\eta_p \operatorname{sgn}(\mu_p) \operatorname{sech}[2\eta_p(x-x_p^+-v_p t)]\right)\right| = 0,$$

$$\text{(2.3b)} \qquad x_p^+ = \frac{1}{2\eta_p} \log\left\{\frac{|\mu_p|}{2\eta_p} \prod_{\ell=1}^{p-1} \left(\frac{\eta_\ell - \eta_p}{\eta_\ell + \eta_p}\right)^2\right\}, \qquad v_p = -\frac{1}{4\eta_p^2}.$$

Note that v_p has negative sign. Thus, all sech-shaped solitons propagate to the left.

In $M > 0$ in Chapter 4, (2.18), then the structure of $q_d(x,t)$ is more complicated. Let us consider the simplest case: $N = 2$, $\zeta_1 = \xi + i\eta$, $\zeta_2 = -\xi + i\eta = -\zeta_1^*$, with $\xi, \eta > 0$. Then $C_1^r = \lambda + i\mu$ and $C_2^r = -\lambda + i\mu$, where λ and μ are real constants not vanishing simultaneously. Using (2.1) one gets (cf. [10] (or Chapter 5), (2.4))

$$\text{(2.4)} \qquad q_d(x,t) = 4\eta \operatorname{sech} \Psi \left[\frac{\sin \Phi + (\eta/\xi)\cos \Phi \tanh \Psi}{1 + (\eta/\xi)^2 \cos^2\Phi \operatorname{sech}^2 \Psi}\right],$$

with

$$\text{(2.5a)} \qquad \Phi = 2\xi x - \xi(2(\xi^2 + \eta^2))^{-1} t + \phi$$

$$\text{(2.5b)} \qquad \Psi = 2\eta x + \eta(2(\xi^2 + \eta^2))^{-1} t + \psi,$$

where the constants ϕ and ψ satisfy

$$\text{(2.6a)} \qquad \exp(-\psi) = \tfrac{1}{2}(\xi/\eta)(\lambda^2 + \mu^2)^{\frac{1}{2}}(\xi^2 + \eta^2)^{-\frac{1}{2}}$$

$$\text{(2.6b)} \qquad \sin \phi = (-\lambda\eta + \mu\xi)(\lambda^2 + \mu^2)^{-\frac{1}{2}}(\xi^2 + \eta^2)^{-\frac{1}{2}}$$

$$\text{(2.6c)} \qquad \cos \phi = (\lambda\xi + \mu\eta)(\lambda^2 + \mu^2)^{-\frac{1}{2}}(\xi^2 + \eta^2)^{-\frac{1}{2}}.$$

Thus $q_d(x,t)$ has the form of a breather (see [6]) with envelope and phase velocities $v_e = (-4(\xi^2 + \eta^2))^{-1}$ and $v_{ph} = (4(\xi^2 + \eta^2))^{-1} = -v_e$, respectively. Observe that the sign of v_e is negative.

If $M > 0$ in Chapter 4, (2.18) and $N > 2$, then, generically, $q_d(x,t)$ will decompose into M breathers and (N - 2M) sech-shaped solitons. However, it is easy to construct examples (e.g. $N = 3$, $\zeta_1 = \xi + i\eta$, $\zeta_3 = i(\xi^2 + \eta^2)^{\frac{1}{2}}$), in which no complete decomposition into breathers and sech-shaped solitons takes place.

Since sech-shaped solitons as well as breather envelopes propagate to the left one expects that, regardless of the location of the bound states, for large t the function $q_d(x,t)$ will be concentrated on the negative x-axis. Let us verify this. By Chapter 4, (5.21) we have

$$q_d(x,t) = -2 \operatorname{Im} \sum_{p,j=1}^{N} \beta_{pj}, \tag{2.7}$$

where $B = (\beta_{pj})$ is the inverse of the matrix A given by (2.2).
Since

$$1 - \sum_{\ell,j=1}^{N} \frac{i}{\zeta_\ell+\zeta_p} \beta_{\ell j} = [c_p^r(t)]^{-1} e^{-2i\zeta_p x} \sum_{j=1}^{N} \beta_{pj}, \tag{2.8}$$

we can rewrite (2.7) as

$$q_d(x,t) = -2 \operatorname{Im} \sum_{p=1}^{N} c_p^r(t) e^{2i\zeta_p x}\left(1 - \sum_{\ell,j=1}^{N} \frac{i}{\zeta_\ell+\zeta_p} \beta_{\ell j}\right). \tag{2.9}$$

Hence, using Chapter 4, (5.13), one gets

$$|q_d(x,t)| \le 2 \sum_{p=1}^{N} |c_p^r|\left(1 + \sum_{\ell,j=1}^{N} |\zeta_\ell+\zeta_p|^{-1} N_{\ell j}\right)\exp\{-\psi_p(x,t)\} \tag{2.10a}$$

$$\psi_p(x,t) = 2(\operatorname{Im} \zeta_p)(x + |2\zeta_p|^{-2}t), \tag{2.10b}$$

where the constants $N_{\ell j}$, introduced in Chapter 4, (5.7b), reappear in this paper in (3.4b). In particular, this shows that

$$\sup_{x>0} |q_d(x,t)| = O(e^{-\omega t}) \quad \text{as } t \to \infty \tag{2.11a}$$

$$\int_0^\infty q_d^2(x,t)dx = O(e^{-2\omega t}) \quad \text{as } t \to \infty, \tag{2.11b}$$

where ω is the positive constant

$$\omega = \min\left\{2(\operatorname{Im} \zeta_p)|2\zeta_p|^{-2};\ p = 1,2,\ldots,N\right\}. \tag{2.12}$$

Thus, indeed, we conclude that, as time goes on, $q_d(x,t)$ is confined to the negative x-axis.

Since $q_d(x,t)$ satisfies Chapter 4, (2.12), we obtain from (2.11b)

$$(2.13)\qquad \int_{-\infty}^{0} q_d^2(x,t)dx = 4\sum_{p=1}^{N} \mathrm{Im}\,\zeta_p + O(e^{-2\omega t}) \qquad \text{as } t \to \infty.$$

3. A useful result obtained earlier.

The fact, that for each $t > 0$ the solution $q(x,t)$ of (1.1) satisfies the hypotheses (2.2-13) in Chapter 4, required for a bona fide potential in the Zakharov-Shabat scattering problem, immediately yields the following result, which is established in Chapter 4 in the form of theorem 4.1 and its corollary.

Theorem 3.1. *Let* $q(x,t)$ *be the solution of the sine-Gordon problem*

$$(3.1)\qquad \begin{cases} q_t = \tfrac{1}{2}\sin\left[2\displaystyle\int_{-\infty}^{x} q(x',t)dx'\right], & -\infty < x < +\infty, \quad t > 0 \\ q(x,0) = q_0(x), \end{cases}$$

where the initial function $q_0(x)$ *is an arbitrary real function on* $\mathbb{R}$, *satisfying (1.2a-b-c(i)). Let* $\{b_r(\zeta),\zeta_j,C_j^r\}$ *be the scattering data associated with* $q_0(x)$. *Then for each* $x \in \mathbb{R}$ *and* $t > 0$ *one has*

$$(3.2)\qquad |q(x,t)-q_d(x,t)| \le a_0^2\left(\int_0^{\infty} |\Omega_c(x+y;t)|^2 dy + \sup_{0<y<+\infty} |\Omega_c(x+y;t)|\right),$$

where $q_d(x,t)$ *is the reflectionless part of* $q(x,t)$ *given by (2.1) and*

$$(3.3)\qquad \Omega_c(s;t) = \frac{1}{\pi}\int_{-\infty}^{\infty} b_r(\zeta)e^{2i\zeta s-it/(2\zeta)}d\zeta, \qquad s \in \mathbb{R},$$

$$(3.4a)\qquad a_0 = 1 + \sum_{p,j=1}^{N} (\mathrm{Im}\,\zeta_p)^{-1}N_{pj},$$

$$(3.4b)\qquad N_{pj} = 2(\mathrm{Im}\,\zeta_p)^{\frac{1}{2}}(\mathrm{Im}\,\zeta_j)^{\frac{1}{2}}\prod_{\substack{\ell=1\\ \ell\neq p}}^{N}\left|\frac{\zeta_p-\zeta_\ell^*}{\zeta_p-\zeta_\ell}\right|\prod_{\substack{k=1\\ k\neq j}}^{N}\left|\frac{\zeta_j-\zeta_k^*}{\zeta_j-\zeta_k}\right| .$$

Furthermore, the following a priori bound holds

(3.5) $$\sup_{(x,t)\in\mathbb{R}\times[0,\infty)} |q(x,t)-q_d(x,t)| \leq \frac{a_0^2}{\pi} \int_{-\infty}^{\infty} \Big(|b_r(\zeta)|+|b_r(\zeta)|^2\Big)d\zeta.$$

4. Estimates of $\Omega_c(x+y;t)$.

Since in (3.2) the quantity a_0 is invariant with time, the magnitude of $q(x,t) - q_d(x,t)$ in the region (1.10) depends only on the behaviour of $\Omega_c(x+y;t)$. This function in turn can be estimated in a simple and explicit way, as is shown in the next lemma.

Lemma 4.1. *In the situation of theorem 3.1, assume that*

(4.1) *There is an integer* $n \geq 2$ *such that* b_r *is of class* $C^n(\mathbb{R})$ *and all derivatives* $b_r^{(j)}(\zeta)$, $j = 0,1,\dots,n$ *satisfy*

$$b_r^{(j)}(\zeta) = O(\zeta^{-2n+2}), \qquad \zeta \to \pm\infty.$$

Let $t > 0$, $y > 0$ *and*

(4.2) $x \geq -\mu - \nu t^\delta$, *where* μ, ν *and* δ *are nonnegative constants.*

Put

(4.3) $$\tau = 1 + \mu + \nu t^\delta, \quad w = x + y + \tau, \quad B(\zeta,t) = b_r(\zeta)e^{-2i\zeta\tau}, \quad B^{(m)} = \left(\frac{\partial}{\partial\zeta}\right)^m B.$$

Then, for fixed n *in (4.1) one has*

(4.4a) $$|\Omega_c(x+y;t)| \leq c_n\tau^n t^{-(2n-3)/2} w^{-3/2},$$

(4.4b) $$c_n = d_n \max\{\|\psi_n b_r^{(j)}\|_\infty;\ j = 0,1,\dots,n\},$$

where d_n *is a constant independent of* b_r, μ, ν *and* δ *and where* $\psi_n \in C(\mathbb{R})$ *is defined by*

(4.4c) $$\psi_n(\zeta) = \max(1,\zeta^{2n-2}), \qquad \zeta \in \mathbb{R}.$$

Proof: For functions $f(x,y,t,\zeta)$ partial derivatives with respect to ζ will be denoted by

(4.5) $$f^{(m)} = (\frac{\partial}{\partial\zeta})^m f.$$

Let us write

(4.6) $$\phi = 2\zeta w - \frac{t}{2\zeta}$$

(4.7) $$s = \frac{1}{\phi^{(1)}} = \frac{2\zeta^2}{4\zeta^2 w+t} \geq 0.$$

Using (4.3-6) and the symmetry relation $b_r^*(\zeta) = b_r(-\zeta)$ (see Chapter 4, (2.24a)), we can rewrite Ω_c as

(4.8) $$\Omega_c(x+y;t) = \frac{2}{\pi} \operatorname{Re} \int_0^\infty e^{i\phi} B d\zeta.$$

Evidently, B belongs to C^n $(-\infty < \zeta < +\infty)$ and by Leibniz' formula

(4.9a) $$|\psi_n(\zeta)B^{(m)}(\zeta,t)| \leq N_{n,m}\tau^m, \text{ with}$$

(4.9b) $$N_{n,m} = 2^{2m}\max\{\|\psi_n b_r^{(j)}\|_\infty;\ j = 0,1,\ldots,m\}.$$

Now, let $0 < r < R < +\infty$. Integrating by parts n times we find

(4.10) $$\int_r^R e^{i\phi}Bd\zeta = -ise^{i\phi}\sum_{\ell=0}^{n-1}(iT)^\ell B\Big|_r^R + \int_r^R e^{i\phi}(iT)^n Bd\zeta,$$

where the operator T is defined by

(4.11) $$Tf = (sf)^{(1)} = s^{(1)}f + sf^{(1)}.$$

Induction reveals that the ℓ-th iterate of T has the structure

(4.12a) $$T^\ell f = s^\ell \sum_{p=0}^{\ell} \alpha_{\ell,p} f^{(\ell-p)} \text{ with } \alpha_{\ell,0} = 1, \text{ whereas for } p \geq 1$$

(4.12b) $$\alpha_{\ell,p} = \sum_{\substack{0\leq\ell_1,\ell_2,\ldots,\ell_p\in\mathbb{Z} \\ \ell_1+2\ell_2+\ldots+p\ell_p=p}} a_{\ell;\ell_1\ell_2\ldots\ell_p}\left(\frac{s^{(1)}}{s}\right)^{\ell_1}\left(\frac{s^{(2)}}{s}\right)^{\ell_2}\cdots\left(\frac{s^{(p)}}{s}\right)^{\ell_p},$$

where $a_{\ell;\ell_1\ell_2\ldots\ell_p}$ are nonnegative integers, independent of s and f. Applying Leibniz' formula to the identity $(4\zeta^2 w+t)s = 2\zeta^2$, we find

(4.13) $$(4\zeta^2 w+t)s^{(j)} + 8j\zeta ws^{(j-1)} + 4j(j-1)ws^{(j-2)} = (2\zeta^2)^{(j)},$$

from which it is easily seen that

$$(4.14)\qquad \left|\frac{s^{(j)}}{s}\right| \le \frac{M_j}{|\zeta|^j},\qquad \zeta \in \mathbb{R}\setminus\{0\},\qquad j = 1,2,\ldots,$$

where M_j is a constant, independent of μ, ν and δ.
Thus, in view of (4.12) there are constants $A_{\ell,p}$, independent of μ, ν and δ, such that for every f of class C^n $(-\infty < \zeta < +\infty)$

$$(4.15)\qquad |T^{\ell} f| \le s^{\ell} \sum_{p=0}^{\ell} \frac{A_{\ell,p}}{|\zeta|^p} |f^{(\ell-p)}|,\qquad \zeta \in \mathbb{R}\setminus\{0\},\qquad \ell = 0,1,\ldots,n.$$

Taking $r \downarrow 0$ and $R \to \infty$ in (4.10) we obtain from (4.8-9-15)

$$(4.16)\qquad \Omega_c(x+y;t) = \frac{2}{\pi} \operatorname{Re} \int_0^{\infty} e^{i\phi} (iT)^n B\, d\zeta,$$

where the integral is absolutely convergent.
Note that by (4.7-9-15)

$$(4.17)\qquad |T^n B| \le s^n \zeta^{2-2n} \sum_{p=0}^{n} A_{n,p} |\psi_n(\zeta) B^{(n-p)}| \le$$

$$\le \frac{2^n \zeta^2}{(4\zeta^2 w+t)^n} N_{n,n} \tau^n \sum_{p=0}^{n} A_{n,p}.$$

Consequently

$$(4.18a)\qquad |\Omega_c(x+y;t)| \le c_n \tau^n t^{-(2n-3)/2} w^{-3/2},\ \text{with}$$

$$(4.18b)\qquad c_n = d_n 2^{-2n} N_{n,n},\qquad d_n = \frac{1}{\pi} 2^{3n-3} \mathcal{B}(\tfrac{3}{2}, n-\tfrac{3}{2}) \sum_{p=0}^{n} A_{n,p},$$

where $\mathcal{B}$ refers to the beta function ([4], p. 258).
Herewith, the proof of the lemma is completed. □

Let the conditions of the preceding lemma be fulfilled. Having fixed n in (4.1) we select λ such that

$$(4.19)\qquad 0 < \lambda \le n - \frac{3}{2}.$$

Next, we choose δ such that

$$(4.20)\qquad 0 \le \delta \le 1 - \left(\frac{3+2\lambda}{2n}\right).$$

By virtue of lemma 4.1 we then have in the parameter region

(4.21) $t \geq 1,\ x \geq -\mu - \nu t^{\delta}$, where μ and ν are nonnegative constants,

the following estimates

(4.22a) $$\sup_{0<y<+\infty} |\Omega_c(x+y;t)| \leq \rho_n t^{-\lambda}(x+\tau)^{-3/2}$$

(4.22b) $$\int_0^{\infty} |\Omega_c(x+y;t)|^2 dy \leq \tfrac{1}{2}\rho_n^2 t^{-2\lambda}(x+\tau)^{-2}, \text{ with}$$

(4.23) $$\rho_n = (1+\mu+\nu)^n c_n .$$

Here $\tau = 1 + \mu + \nu t^{\delta}$, while c_n denotes the constant introduced in (4.4).

5. Estimates of $q(x,t) - q_d(x,t)$ and $q(x,t)$.

It is now merely a matter of combining theorem 3.1 with the estimates of $\Omega_c(x+y;t)$ derived in section 4 to obtain the main result of this chapter, which we state as follows

Theorem 5.1. *Let* $q(x,t)$ *be the solution of the sine-Gordon problem*

(5.1) $$\begin{cases} q_t = \tfrac{1}{2}\sin\left[2\int_{-\infty}^{x} q(x',t)dx'\right], & -\infty < x < +\infty, \quad t > 0 \\ q(x,0) = q_0(x), \end{cases}$$

where the initial function $q_0(x)$ *is an arbitrary real function on* $\mathbb{R}$, *satisfying (1.2) in such a way that (4.1) is fulfilled. Let* $\{b_r(\zeta),\zeta_j,C_j^r\}$ *be the scattering data associated with* $q_0(x)$. *Then for each* $x \in \mathbb{R}$ *and* $t > 0$ *one has*

(5.2) $$|q(x,t) - q_d(x,t)| \leq a_0^2\left(\int_0^{\infty} |\Omega_c(x+y;t)|^2 dy + \sup_{0<y<+\infty} |\Omega_c(x+y;t)|\right),$$

with $q_d(x,t)$ *the reflectionless part (2.1) of* $q(x,t)$, a_0 *the constant given by (3.4) and* Ω_c *the function introduced in (3.3).*

Next, fix n *in (4.1). Let the constants* μ, ν, λ *and* δ *satisfy*

(5.3) $$\mu \geq 0, \quad \nu \geq 0, \quad 0 < \lambda \leq n - \frac{3}{2}, \quad 0 \leq \delta \leq 1 - \left(\frac{3+2\lambda}{2n}\right).$$

Put $\alpha = \mu + \nu t^{\delta}$. *Then one has the estimates*

$$(5.4a)\qquad |q(x,t) - q_d(x,t)| \leq A \quad \textit{for } t > 0, \quad x \geq -\alpha$$

$$(5.4b)\qquad |q(x,t) - q_d(x,t)| \leq \tilde{\rho}_n t^{-\lambda}(x + 1 + \alpha)^{-3/2} \quad \textit{for } t \geq 1,\ x \geq -\alpha$$

with

$$(5.5a)\qquad A = \frac{a_0^2}{\pi} \int_{-\infty}^{\infty} (|b_r(\zeta)| + |b_r(\zeta)|^2)d\zeta$$

$$(5.5b)\qquad \tilde{\rho}_n = a_0^2\rho_n(1 + \tfrac{1}{2}\rho_n),$$

where the constant ρ_n *is given by (4.23).*

To find the behaviour of $q(x,t)$ in the coordinate region $t > 0$, $x \geq -\alpha$, it clearly suffices to combine the preceding theorem with the results of section 2.

It follows from the last remark of Chapter 4, section 6, that the estimates (5.2-4) still hold if one replaces the left hand side by

$$(5.6)\qquad \left|\int_{x}^{\infty} (q^2(x',t) - q_d^2(x',t))dx'\right|.$$

In particular, this yields

$$(5.7)\qquad \int_{-\alpha}^{\infty} q^2(x,t)dx = \int_{-\alpha}^{\infty} q_d^2(x,t)dx + O(t^{-\lambda}) \quad \text{as } t \to \infty.$$

However, by (2.10) one has

$$(5.8)\qquad \int_{-\alpha}^{\infty} q_d^2(x,t)dx = O(e^{-2\tilde{\omega}t}) \quad \text{as } t \to \infty$$

where $\tilde{\omega}$ is any constant satisfying $0 < \tilde{\omega} < \omega$ with ω as in (2.12). Consequently

$$(5.9a)\qquad \int_{-\alpha}^{\infty} q^2(x,t)dx = O(t^{-\lambda}) \quad \text{as } t \to \infty.$$

Using the formula (see [3])

$$(5.10)\qquad \int_{-\infty}^{\infty} q^2(x,t)dx = \frac{2}{\pi}\int_{0}^{\infty} \log(1 + |b_r(\zeta)|^2)d\zeta + 4\sum_{p=1}^{N} \operatorname{Im} \zeta_p,$$

we obtain for the complementary integral

(5.9b) $$\int_{-\infty}^{-\alpha} q^2(x,t)dx = \frac{2}{\pi}\int_0^{\infty} \log(1+|b_r(\zeta)|^2)d\zeta + 4\sum_{p=1}^{N} \operatorname{Im}\zeta_p + O(t^{-\lambda})$$

as $t \to \infty$.

Note that in addition to (5.8) it also follows from (2.10) that

(5.11) $$\sup_{x\geq-\alpha} |q_d(x,t)| = O(e^{-\tilde{\omega}t}) \quad \text{as } t \to \infty.$$

Thus, in view of (5.4), we arrive at

(5.12) $$\sup_{x\geq-\alpha} |q(x,t)| = O(t^{-\lambda}) \quad \text{as } t \to \infty.$$

Let us mention some consequences of the above results for the solution $\sigma(x,t)$, given by (1.5), of the sine-Gordon problem (1.6). By (1.3) and (5.4b) one has for $t \geq 1$, $x \geq -\alpha$

(5.13) $$|\sigma(x,t) - \sigma_d(x,t) + 2(k_0-k_1)\pi| \leq 4\tilde{\rho}_n t^{-\lambda}(x+1+\alpha)^{-\frac{1}{2}},$$ with

(5.14a) $$\sigma_d(x,t) = -2\int_{-\infty}^{x} q_d(x',t)dx'$$ and

(5.14b) $$k_1 = \frac{1}{\pi}\int_{-\infty}^{\infty} q_d(x,t)dx \in \mathbb{Z}.$$

Using (2.10), we obtain

(5.15) $$\sup_{x\geq-\alpha} |\sigma(x,t) + 2k_0\pi| = O(t^{-\lambda}) \quad \text{as } t \to \infty.$$

To conclude with, let us observe that if the initial function $q_0(x)$ in (1.1b) is in the Schwartz class, then so is b_r.
This implies that (4.1) holds for all n. Hence (5.7-9-12-15) are in this case valid for all $\lambda > 0$ and all $0 \leq \delta < 1$.

References

[1] M.J. Ablowitz, D.J. Kaup, A.C. Newell and H. Segur, Method for solving the sine-Gordon equation, Phys. Rev. Lett. 30 (1973), 1262-1264.

[2] M.J. Ablowitz, D.J. Kaup, A.C. Newell and H. Segur, The inverse scattering transform - Fourier analysis for nonlinear problems, Stud. Appl. Math. 53 (1974), 249-315.

[3] M.J. Ablowitz and H. Segur, Solitons and the Inverse Scattering Transform, Philadelphia, SIAM, 1981.

[4] M. Abramowitz and I.A. Stegun, Handbook of Mathematical Functions, National Bureau of Standards Applied Mathematics Series, No. 55, U.S. Department of Commerce, 1964.

[5] D.J. Kaup and A.C. Newell, The Goursat and Cauchy problems for the sine-Gordon equation, SIAM J. Appl. Math. 34 (1978), 37-54.

[6] G.L. Lamb, Jr., Elements of Soliton Theory, Wiley-Interscience, 1980.

[7] G.L. Lamb, Jr. and D.W. McLaughlin, Aspects of soliton physics, in: Solitons (Ed. R.K. Bullough and P.J. Caudrey) Topics in Current Physics 17, Springer-Verlag, New York, 1980.

[8] M. Ohmiya, On the generalized soliton solutions of the modified Korteweg-de Vries equation, Osaka J. Math. 11 (1974), 61-71.

[9] P. Schuur, On the approximation of a real potential in the Zakharov-Shabat system by its reflectionless part, preprint 341, Mathematical Institute Utrecht (1984).

[10] P. Schuur, Decomposition and estimates of solutions of the modified Korteweg-de Vries equation on right half lines slowly moving leftward, preprint 342, Mathematical Institute Utrecht (1984).

[11] S. Tanaka, Non-linear Schrödinger equation and modified Korteweg-de Vries equation; construction of solutions in terms of scattering data, Publ. R.I.M.S. Kyoto Univ. 10 (1975), 329-357.

CHAPTER EIGHT

ON THE APPROXIMATION OF A COMPLEX POTENTIAL IN THE ZAKHAROV-SHABAT SYSTEM BY ITS REFLECTIONLESS PART

We consider the Zakharov-Shabat system with complex potential and derive a pointwise estimate of the error made in approximating the potential by its reflectionless part. As an illustration we apply this estimate to investigate the long-time behaviour of the solution of the complex modified Korteweg-de Vries initial value problem.

1. Introduction.

In [8], i.e. Chapter 4 of this volume, studying the Zakharov-Shabat system with real potential, we derived a pointwise estimate of the error made in approximating the potential by its reflectionless part. For that purpose we reduced the matrix Gel'fand-Levitan equation appearing in the literature to a scalar integral equation containing only a single integral. Furthermore, we exploited the fact that the corresponding scalar Gel'fand-Levitan operator, when considered in the complex Hilbert space $L^2(0,\infty)$, has the structure of the identity plus an antisymmetric operator.

In this chapter we take a more general standpoint by considering the Zakharov-Shabat system with a potential that may assume complex values. This confronts us with a matrix Gel'fand-Levitan equation that can no longer be reduced in the above way. It has, however, one important property in common with the scalar case: The corresponding matrix Gel'fand-Levitan operator has, when considered in the complex Hilbert space $(L^2(0,\infty))^{2\times 2}$, the structure of the identity plus an antisymmetric operator.

Motivated by this resemblance we present an analysis of the matrix Gel'fand-Levitan equation that - as far as its abstract setting is concerned - parallels the analysis of the scalar Gel'fand-Levitan equation given in Chapter 4. To increase the similarity matrix norms and notation are selected with due care. Although the technicalities are different (cf. the proof of lemma 5.1 with that of Chapter 4, lemma 5.1) this analysis leads to an estimate of the difference of the potential and its reflectionless part, which, surprisingly enough, is almost identical with that obtained in Chapter 4. Since the only alteration consists in some numerical front factors due to the particular matrix norms involved, we may truly speak of a generalization of the main result of Chapter 4.

This generalization is of practical importance, since working with a complex instead of a real potential considerably enlarges the class of nonlinear evolution equations solvable via the associated inverse scattering method. For example, the complex modified Korteweg-de Vries equation, as well as the nonlinear Schrödinger equation can be solved by the complex but not by the real Zakharov-Shabat inverse scattering method.

The composition of the chapter is as follows.

In section 2 we briefly discuss the direct scattering problem for the Zakharov-Shabat system with complex potential. The inverse scattering formalism is outlined in section 3. Next, in section 4 we state our main result, which after the introduction of some convenient notation and the derivation of a useful lemma in section 5, is proven in section 6. Finally we apply the aforementioned result to investigate the long-time behaviour of the solution of the complex modified Korteweg-de Vries initial value problem.

2. Direct scattering.

To begin with let us tersely survey the direct scattering problem for the Zakharov-Shabat system [11]

$$(2.1)\qquad \begin{pmatrix}\psi_1\\ \psi_2\end{pmatrix}' = \begin{pmatrix}-i\zeta & q\\ -q^* & i\zeta\end{pmatrix}\begin{pmatrix}\psi_1\\ \psi_2\end{pmatrix}, \qquad ' = \frac{d}{dx}, \qquad -\infty < x < +\infty$$

where $q = q(x)$ is a complex function and ζ a complex parameter. For details and proofs we refer to [1], [2], [5], [10]. Our notation is similar to that used in [5].

Following [5] we assume throughout that the potential q has the regularity and decay properties stated below:

$$(2.2a)\qquad q \in C^1(\mathbb{R})$$

$$(2.2b)\qquad \lim_{|x|\to\infty} q(x) = \lim_{|x|\to\infty} q'(x) = 0$$

$$(2.2c)\qquad \int_{-\infty}^{\infty} (|q(s)| + |q'(s)|)ds < +\infty.$$

In addition we shall require some conditions on the zeros of the Wronskian of the right and left Jost solutions, to be specified presently in (2.11).

It is interesting to compare the results of this section to those of Chapter 4, section 2 and to single out the symmetry relations in Chapter 4 caused by the realness of the potential.

For $\mathrm{Im}\,\zeta \geq 0$ we define the (right and left) Jost solutions $\psi_r(x,\zeta)$ and $\psi_\ell(x,\zeta)$ as the special solutions of (2.1) uniquely determined by

$$(2.3a)\qquad \psi_r(x,\zeta) = e^{-i\zeta x}R(x,\zeta), \qquad \lim_{x\to-\infty} R(x,\zeta) = \binom{1}{0}$$

$$(2.3b)\qquad \psi_\ell(x,\zeta) = e^{i\zeta x}L(x,\zeta), \qquad \lim_{x\to+\infty} L(x,\zeta) = \binom{0}{1}.$$

The vector functions R and L are continuous in (x,ζ) on $\mathbb{R}\times\overline{\mathbb{C}}_+$ and analytic in ζ on $\mathbb{C}_+$ for each $x \in \mathbb{R}$.

Furthermore, their components satisfy

$$(2.4)\qquad \max\Big[\sup_{\mathbb{R}\times\overline{\mathbb{C}}_+} |R_i(x,\zeta)|,\ \sup_{\mathbb{R}\times\mathbb{C}_+} |L_i(x,\zeta)|\Big] \leq \exp\Big\{\int_{-\infty}^{\infty} |q(s)|ds\Big\},$$

$$i = 1,2.$$

For Im $\zeta \leq 0$ we set

$$(2.5)\qquad \tilde{\psi}_r(x,\zeta) \equiv \begin{pmatrix} -\psi^*_{r_2}(x,\zeta^*) \\ \psi^*_{r_1}(x,\zeta^*) \end{pmatrix}, \qquad \tilde{\psi}_\ell(x,\zeta) \equiv \begin{pmatrix} \psi^*_{\ell_2}(x,\zeta^*) \\ -\psi^*_{\ell_1}(x,\zeta^*) \end{pmatrix}.$$

It is readily verified that $\tilde{\psi}_r$ and $\tilde{\psi}_\ell$ are solutions of (2.1). Moreover, for x, $\zeta \in \mathbb{R}$ one has

$$(2.6a)\qquad W(\psi_r,\tilde{\psi}_r) = |R_1(x,\zeta)|^2 + |R_2(x,\zeta)|^2 = 1$$

$$(2.6b)\qquad W(\tilde{\psi}_\ell,\psi_\ell) = |L_1(x,\zeta)|^2 + |L_2(x,\zeta)|^2 = 1 ,$$

where $W(\psi,\phi) = \psi_1\phi_2 - \psi_2\phi_1$ denotes the Wronskian of ψ and ϕ. Hence, for ζ real, the pairs $\psi_r,\tilde{\psi}_r$ and $\psi_\ell,\tilde{\psi}_\ell$ constitute fundamental systems of solutions of equation (2.1). In particular, we have for x, $\zeta \in \mathbb{R}$

$$(2.7a)\qquad \psi_r(x,\zeta) = r_+(\zeta)\psi_\ell(x,\zeta) + r_-(\zeta)\tilde{\psi}_\ell(x,\zeta)$$

$$(2.7b)\qquad r_+(\zeta) = W(\tilde{\psi}_\ell,\psi_r)$$

$$(2.7c)\qquad r_-(\zeta) = W(\psi_r,\psi_\ell).$$

It is not hard to show that

$$(2.8)\qquad |r_+(\zeta)|^2 + |r_-(\zeta)|^2 = 1, \qquad \zeta \in \mathbb{R}.$$

Let us use (2.7c) to extend $r_-(\zeta)$ to a function analytic on Im $\zeta > 0$ and continuous on Im $\zeta \geq 0$.
Then the following integral representations hold

$$(2.9a)\qquad r_+(\zeta) = -\int_{-\infty}^{\infty} q^*(s)e^{-2i\zeta s}R_1(s,\zeta)ds, \qquad \zeta \in \mathbb{R}$$

$$(2.9b)\qquad r_-(\zeta) = 1 + \int_{-\infty}^{\infty} q(s)R_2(s,\zeta)ds, \qquad \text{Im}\,\zeta \geq 0.$$

In combination with (2.6a) these yield

$$(2.10)\qquad \max\left[\sup_{\zeta\in\mathbb{R}} |r_+(\zeta)|,\ \sup_{\zeta\in\mathbb{R}} |1 - r_-(\zeta)|\right] \leq \int_{-\infty}^{\infty} |q(s)|ds.$$

In terms of r_- we make our final assumptions:

$$(2.11a)\qquad r_-(\zeta) \neq 0 \text{ for } \zeta \in \mathbb{R}$$

(2.11b) All zeros of r_- in $\mathbb{C}_+$ are simple.

Let us point out that condition (2.11b) can be circumvented by using Tanaka's direct and inverse scattering formalism [10]. We only include it for reasons of simplicity.
Incidentally, if

$$\int_{-\infty}^{\infty} |q(s)|ds < 1 \tag{2.12}$$

then (2.10) shows that (2.11a) is fulfilled.
Moreover, if

$$\int_{-\infty}^{\infty} |q(s)|ds < 0.904 \tag{2.13}$$

then (2.11a) and (2.11b) are trivially fulfilled since $r_-(\zeta) \neq 0$ for $\operatorname{Im} \zeta \geq 0$ (see [2]).

We now turn to the construction of the scattering data associated with q(x). As a result of (2.11a) the function $r_-(\zeta)$ has at most finitely many zeros $\zeta_1,\zeta_2,\ldots,\zeta_N$, $\operatorname{Im} \zeta_j > 0$. They are all simple by virtue of (2.11b). It is a remarkable fact, that the ζ_j are precisely the eigenvalues of (2.1) in the upper half plane (the so-called bound states). The associated L^2-eigenspaces are one-dimensional and spanned by the exponentially decaying vector functions $\psi_\ell(x,\zeta_j)$, $j = 1,2,\ldots,N$. Note that by (2.7c) there are nonzero constants $\alpha(\zeta_j)$ such that

$$\psi_r(x,\zeta_j) = \alpha(\zeta_j)\psi_\ell(x,\zeta_j). \tag{2.14}$$

One can derive the following representation

$$\frac{dr_-}{d\zeta}(\zeta_j) = -2i\alpha(\zeta_j) \int_{-\infty}^{\infty} \psi_{\ell_1}(s,\zeta_j)\psi_{\ell_2}(s,\zeta_j)ds. \tag{2.15}$$

Bearing in mind that the integral on the right does not vanish because of (2.11b) we define the (right) normalization coefficients by

$$c_j^r = \tfrac{1}{2}i\left[\int_{-\infty}^{\infty} \psi_{\ell_1}(s,\zeta_j)\psi_{\ell_2}(s,\zeta_j)ds\right]^{-1}. \tag{2.16}$$

Next, we introduce the following functions of $\zeta \in \mathbb{R}$

(2.17a) $a_r(\zeta) = 1/r_-(\zeta)$, the (right) transmission coefficient

(2.17b) $\quad b_r(\zeta) = r_+(\zeta)/r_-(\zeta)$, the (right) reflection coefficient.

By (2.8) one has for $\zeta \in \mathbb{R}$

(2.18) $$|a_r(\zeta)|^2 - |b_r(\zeta)|^2 = 1.$$

In [5] it is shown that b_r is an element of $C \cap L^1 \cap L^2(\mathbb{R})$, which behaves as $o(|\zeta|^{-1})$ for $\zeta \to \pm\infty$. Of course, by imposing stronger regularity and decay conditions on $q(x)$ in addition to (2.2-11), one can improve the behaviour of $b_r(\zeta)$. For instance, if $q(x)$ has rapidly decaying derivatives, then so has $b_r(\zeta)$.
We shall call the aggregate of quantities $\{b_r(\zeta),\zeta_j,C_j^r\}$ the (right) scattering data associated with the potential q. Remarkably enough, a potential is completely determined by its scattering data.

3. Inverse scattering.

Let q be any potential satisfying (2.2-11). Then q can be recovered from its scattering data $\{b_r(\zeta),\zeta_j,C_j^r\}$ by solving the inverse scattering problem.
For that purpose one defines the following functions of $s \in \mathbb{R}$

(3.1a) $$\Omega(s) = \Omega_d(s) + \Omega_c(s),$$

(3.1b) $$\Omega_d(s) = -2i \sum_{j=1}^{N} C_j^r e^{2i\zeta_j s},$$

(3.1c) $$\Omega_c(s) = \frac{1}{\pi} \int_{-\infty}^{\infty} b_r(\zeta) e^{2i\zeta s} d\zeta.$$

Since b_r is in $C_0 \cap L^1(\mathbb{R})$, the integral in (3.1c) converges absolutely and Ω_c belongs to $C_0 \cap L^2(\mathbb{R})$.
Next, introduce the 2×2 matrix

(3.2) $$\omega(s) = \begin{pmatrix} 0 & -\Omega^*(s) \\ \Omega(s) & 0 \end{pmatrix}$$

and consider the Gel'fand-Levitan equation (see [1], [2], [5], [10])

$$(3.3)\qquad \beta(y;x) + \omega(x+y) + \int_0^\infty \beta(z;x)\omega(x+y+z)dz = 0$$

with $y > 0$, $x \in \mathbb{R}$. In this integral equation the unknown $\beta(y;x)$ is a 2×2 matrix function of the variable y, whereas x is a parameter. Observe that some authors use a slightly different version of the Gel'fand-Levitan equation which can be transformed into (3.3) by a change of variables (see [5], p. 46).

In [2] it is shown that for each $x \in \mathbb{R}$ there is a unique solution $\beta(y;x)$ to (3.3) in $(L^2)^{2\times 2}$ $(0 < y < +\infty)$. It has the form

$$(3.4)\qquad \beta = \begin{pmatrix} a^* & -b \\ b^* & a \end{pmatrix},$$

where $a(y;x)$ and $b(y;x)$ are complex functions belonging to $C \cap L^1 \cap L^2$ $(0 < y < +\infty)$, which vanish as $y \to +\infty$. The inverse scattering problem is now solved, since the functions a and b are related to the potential q in the following way

$$(3.5a)\qquad q(x) = b(0^+;x)$$

$$(3.5b)\qquad \int_x^\infty |q(s)|^2 ds = -a(0^+;x), \qquad x \in \mathbb{R}.$$

Using (3.4), the matrix integral equation (3.3) can be reduced to a scalar integral equation involving only b

$$(3.6)\qquad b(y;x) + \Omega^*(x+y) + \int_0^\infty \int_0^\infty b(z;x)\Omega(z+s+x)\Omega^*(s+y+x)dsdz = 0.$$

In this form the Gel'fand-Levitan equation frequently appears in the literature (cf. [1], [10]). However, for our present analysis the matrix form (3.3) proves to be more convenient.

4. Statement of the main result.

If q is a potential with scattering data $\{b_r(\zeta),\zeta_j,C_j^r\}$ then the potential q_d with scattering data $\{0,\zeta_j,C_j^r\}$ is called the reflectionless part of q. The function $q_d(x)$ can be obtained in explicit form (see (5.30)) by solving the Gel'fand-Levitan equation (3.3), which in that case reduces to a system of N linear algebraic equations. The main result of this chapter is the next theorem which tells us in which sense the potential is approximated by its reflectionless part.

Theorem 4.1. *Let* q *be a potential in the Zakharov-Shabat system (2.1), which satisfies (2.2-11) and has the scattering data* $\{b_r(\zeta),\zeta_j,C_j^r\}$. *Let* q_d *denote the reflectionless part of* q. *Then for each* $x \in \mathbb{R}$

$$(4.1) \qquad |q(x) - q_d(x)| \leq \alpha_0^2\left(\sqrt{2}\int_0^{\infty} |\Omega_c(x+y)|^2 dy + \sup_{0<y<+\infty} |\Omega_c(x+y)|\right),$$

with Ω_c *given by (3.1c) and* α_0 *the following explicit function of the bound states* ζ_j

$$(4.2a) \qquad \alpha_0 = 1 + \sum_{p,j=1}^{N} (\operatorname{Im}\zeta_p)^{-1} M_{pj},$$

$$(4.2b) \qquad M_{pj} = 8(\operatorname{Im}\zeta_p)^{\frac{1}{2}}(\operatorname{Im}\zeta_j)^{\frac{1}{2}} \prod_{\substack{\ell=1\\ \ell\neq p}}^{N} \left|\frac{\zeta_p-\zeta_\ell^*}{\zeta_p-\zeta_\ell}\right| \prod_{\substack{k=1\\ k\neq j}}^{N} \left|\frac{\zeta_j-\zeta_k^*}{\zeta_j-\zeta_k}\right| .$$

Herewith, $q-q_d$ *is estimated completely in terms of the scattering data. More precisely, the bound given by (4.1-2) depends only on the reflection coefficient* $b_r(\zeta)$ *and the bound states* ζ_j *and not on the normalization coefficients* C_j^r.

Corollary to theorem 4.1. *Under the conditions of theorem 4.1 we have the a priori bound*

$$(4.3) \qquad \sup_{x\in\mathbb{R}} |q(x) - q_d(x)| \leq \frac{\alpha_0^2}{\pi}\int_{-\infty}^{\infty} (|b_r(\zeta)| + \sqrt{2}|b_r(\zeta)|^2)d\zeta .$$

We shall prove theorem 4.1 in section 6. Before doing so we introduce some notation and derive a useful lemma in section 5.

5. First steps to the proof.

In the remainder of the chapter it is understood that the conditions of theorem 4.1 are fulfilled.

We begin by introducing some useful concepts and notation.

For 2×2 matrices

(5.1) $$A = \begin{pmatrix} \alpha_1 & \alpha_3 \\ \alpha_2 & \alpha_4 \end{pmatrix}, \qquad B = \begin{pmatrix} \beta_1 & \beta_3 \\ \beta_2 & \beta_4 \end{pmatrix}$$

with complex entries we set

(5.2) $$\langle A,B\rangle_E = \sum_{i=1}^{4} \alpha_i \beta_i^*, \qquad \|A\|_E = (\langle A,A\rangle_E)^{\frac{1}{2}}.$$

By $\mathcal{B}$ we denote the linear space of all 2×2 matrix functions

(5.3) $$g(y) = \begin{pmatrix} g_1(y) & g_3(y) \\ g_2(y) & g_4(y) \end{pmatrix},$$

such that each g_i is a complex-valued, continuous and bounded function on $(0,\infty)$. We turn $\mathcal{B}$ into a complex Banach space by equipping it with the norm

(5.4) $$\|g\| = \sup_{0<y<+\infty} \|g(y)\|_E.$$

Furthermore, we write $\mathcal{H}$ to indicate the complex Hilbert space $(L^2(0,\infty))^{2\times 2}$ with inner product

(5.5) $$\langle f,g\rangle = \int_0^\infty \langle f(y),g(y)\rangle_E dy$$

and corresponding norm $\| \ \|_2$.

The choice of the above spaces is motivated by our wish to exploit analogies with the scalar Gel'fand-Levitan equation (3.8) in Chapter 4. Returning to (3.2), let us put

(5.6a) $$\omega(s) = \omega_d(s) + \omega_c(s),$$

(5.6b) $$\omega_d(s) = \begin{pmatrix} 0 & -\Omega_d^*(s) \\ \Omega_d(s) & 0 \end{pmatrix},$$

(5.6c) $$\omega_c(s) = \begin{pmatrix} 0 & -\Omega_c^*(s) \\ \Omega_c(s) & 0 \end{pmatrix}.$$

From section 3 we know that for each $x \in \mathbb{R}$ the functions $y \mapsto \omega_c(x+y)$, $\omega_d(x+y)$ belong to $\mathcal{B} \cap \mathcal{H}$.

Next, keeping $x \in \mathbb{R}$ fixed, we formally write

$$\text{(5.7a)}\qquad (T_d g)(y) = \int_0^\infty g(z)\omega_d(x+y+z)dz$$

$$\text{(5.7b)}\qquad (T_c g)(y) = \int_0^\infty g(z)\omega_c(x+y+z)dz$$

with g as in (5.3). Plainly, T_d can be considered as a mapping from $\mathcal{B}$ into $\mathcal{B}$, but equally well as a mapping from $\mathcal{H}$ into $\mathcal{H}$. On the other hand, T_c is not necessarily a mapping from $\mathcal{B}$ into $\mathcal{B}$. However, suitably modifying formula (4.5.10) in [5] one can easily show that T_c maps $\mathcal{H}$ into $\mathcal{H}$ with a norm that satisfies

$$\text{(5.8)}\qquad \|T_c\|_2 \leq \sup_{\zeta\in\mathbb{R}} |b_r(\zeta)|.$$

It is straightforward to verify that the operators T_d and T_c are both antisymmetric on $\mathcal{H}$, i.e. $T_d^* = -T_d$, $T_c^* = -T_c$. This fact plays a dominating role in our analysis.
In the above abstract language, the Gel'fand-Levitan equation (3.3) takes the form

$$\text{(5.9a)}\qquad (I + T_c + T_d)\beta = -\omega$$

$$\text{(5.9b)}\qquad \omega = \omega_c + \omega_d$$

where I is the identity mapping. A first advantage of this formulation is readily seen. Since $T_c + T_d$ is antisymmetric, the operator $I + T_c + T_d$ is invertible on $\mathcal{H}$ and so we know at once that (5.9) has a unique solution $\beta \in \mathcal{H}$. Note that this fact was already mentioned in section 3, from which we recall that, moreover, $\beta \in \mathcal{B} \cap \mathcal{H}$.

For the proof of theorem 4.1 the following lemma is basic.

Lemma 5.1. *For any value of the parameter* $x \in \mathbb{R}$, *the operator* $I+T_d$ *is invertible on the Banach space* $\mathcal{B}$ *with inverse* $S = (I+T_d)^{-1}$ *given by*

$$\text{(5.10a)}\qquad (Sf)(y) = f(y) - \sum_{p=1}^{N} A_p \begin{pmatrix} 0 & \varepsilon_p^*(y) \\ \varepsilon_p(y) & 0 \end{pmatrix}, \quad \varepsilon_p(y) = e^{2i\zeta_p y},$$

$$\text{(5.10b)}\qquad A_p = \sum_{j=1}^{N} \left(\int_0^\infty f(z) \begin{pmatrix} 0 & \varepsilon_j^*(z) \\ \varepsilon_j(z) & 0 \end{pmatrix} dz\right) \begin{pmatrix} \beta_{p,j} & \beta_{p+N,j} \\ \beta_{p,j+N} & \beta_{p+N,j+N} \end{pmatrix},$$

where (β_{rs}) *is the inverse of the* 2N×2N *matrix*

(5.11a) $$A = \begin{pmatrix} Z & D \\ -D^* & Z^t \end{pmatrix}, \text{ with}$$

(5.11b) $$Z = (\zeta_{pj})_{p,j=1}^{N}, \; \zeta_{pj} = (2i(\zeta_j^* - \zeta_p))^{-1}, \; Z^t = (\zeta_{jp})_{p,j=1}^{N}$$

(5.11c) $$D = (\delta_j \delta_{pj}), \; \delta_j = (-2iC_j^r)^{-1} e^{-2i\zeta_j x}, \; D^* = (\delta_j^* \delta_{pj}).$$

Furthermore, the operator S *satisfies the bound*

(5.12) $$\|S\| \le \alpha_0, \quad x \in \mathbb{R}$$

(5.13a) $$\alpha_0 = 1 + \sum_{p,j=1}^{N} (\mathrm{Im}\,\zeta_p)^{-1} M_{pj}$$

(5.13b) $$M_{pj} = 8(\mathrm{Im}\,\zeta_p)^{\frac{1}{2}} (\mathrm{Im}\,\zeta_j)^{\frac{1}{2}} \prod_{\substack{\ell=1 \\ \ell \ne p}}^{N} \left| \frac{\zeta_p - \zeta_\ell^*}{\zeta_p - \zeta_\ell} \right| \prod_{\substack{k=1 \\ k \ne j}}^{N} \left| \frac{\zeta_j - \zeta_k^*}{\zeta_j - \zeta_k} \right|.$$

Thus, $\|S\|$ *is uniformly bounded for* $x \in \mathbb{R}$, *where the bound is an explicit function of the* ζ_j.

<u>Proof</u>: Let $x \in \mathbb{R}$ be arbitrarily fixed.
Recall that, when considered as an operator from the Hilbert space $\mathcal{H}$ into itself, T_d is antisymmetric. Hence $I+T_d$ is invertible on $\mathcal{H}$ and one has

(5.14) $$\|(I + T_d)g\|_2^2 = \|g\|_2^2 + \|T_d g\|_2^2, \quad g \in \mathcal{H},$$

so that the inverse $S = (I+T_d)^{-1}$ satisfies the bounds

(5.15) $$\|S\|_2 \le 1, \quad \|T_d S\|_2 \le 1.$$

Shifting our gaze, let us consider T_d as an operator from $\mathcal{B}$ into $\mathcal{B}$ and show that $I+T_d$ is invertible on $\mathcal{B}$. Suppose that $(I+T_d)g = 0$ for some $g \in \mathcal{B}$. Then $g = -T_d g \in \mathcal{H}$ and thus g is identically zero by the preceding argument. This tells us that $I+T_d$ is one to one on $\mathcal{B}$. However, T_d is of finite rank and therefore compact. It follows that $I+T_d$ is invertible on the Banach space $\mathcal{B}$.
Next, consider in $\mathcal{B} \cap \mathcal{H}$ the elements $e_1, e_2, \ldots, e_{4N}$ defined by

$$(5.16a)\quad e_j(y) = \begin{pmatrix} 0 & \varepsilon_j^*(y) \\ 0 & 0 \end{pmatrix}, \qquad e_{j+N}(y) = \begin{pmatrix} \varepsilon_j(y) & 0 \\ 0 & 0 \end{pmatrix}$$

$$(5.16b)\quad e_{j+2N}(y) = \begin{pmatrix} 0 & 0 \\ 0 & \varepsilon_j^*(y) \end{pmatrix}, \; e_{j+3N}(y) = \begin{pmatrix} 0 & 0 \\ \varepsilon_j(y) & 0 \end{pmatrix}, \; j=1,2,\ldots,N.$$

Solving the equation

$$(5.17)\quad (I + T_d)g = f, \quad f,g \in \mathcal{B}$$

we find

$$(5.18)\quad g(y) = f(y) - \sum_{p=1}^{N} A_p \begin{pmatrix} 0 & \varepsilon_p^*(y) \\ \varepsilon_p(y) & 0 \end{pmatrix}$$

where the A_p satisfy

$$(5.19a)\quad A_p = \begin{pmatrix} \alpha_p & \alpha_{p+N} \\ \alpha_{p+2N} & \alpha_{p+3N} \end{pmatrix}$$

$$(5.19b)\quad \sum_{s=1}^{2N} \alpha_{rs} \begin{pmatrix} \alpha_s \\ \alpha_{s+2N} \end{pmatrix} = \begin{pmatrix} \beta_r \\ \beta_{r+2N} \end{pmatrix}, \qquad r = 1,2,\ldots,2N$$

$$(5.19c)\quad \beta_q = \int_0^\infty \langle f(y), e_q(y) \rangle_E \, dy, \qquad q = 1,2,\ldots,4N$$

with $A = (\alpha_{rs})$ the matrix given by (5.11).
Since the operator $I+T_d$ is one to one on $\mathcal{B}$, the matrix $A = (\alpha_{rs})$ is invertible. More directly, one can verify the invertibility of A by writing it as a positive definite matrix plus an antisymmetric one. Denoting the inverse by $A^{-1} = (\beta_{rs})$, we obtain from (5.19)

$$(5.20)\quad A_p = \sum_{s=1}^{2N} \begin{pmatrix} \beta_{p,s}\beta_s & \beta_{p+N,s}\beta_s \\ \beta_{p,s}\beta_{s+2N} & \beta_{p+N,s}\beta_{s+2N} \end{pmatrix}$$

$$= \sum_{j=1}^{N} \begin{pmatrix} \beta_j & \beta_{j+N} \\ \beta_{j+2N} & \beta_{j+3N} \end{pmatrix} \begin{pmatrix} \beta_{p,j} & \beta_{p+N,j} \\ \beta_{p,j+N} & \beta_{p+N,j+N} \end{pmatrix}.$$

Together, (5.16), (5.18) and (5.20) imply that the inverse operator $S = (I+T_d)^{-1}$ is given in explicit form by (5.10).

We shall now prove that the matrix elements β_{rs} are bounded as functions of $x \in \mathbb{R}$. In fact, we shall estimate the x dependent matrix that occurs in the right hand side of (5.10b), in the following explicit way

$$(5.21)\qquad \left\|\begin{pmatrix} \beta_{p,j} & \beta_{p+N,j} \\ \beta_{p,j+N} & \beta_{p+N,j+N} \end{pmatrix}\right\|_E \leq 8(\operatorname{Im}\zeta_p)^{\frac{1}{2}}(\operatorname{Im}\zeta_j)^{\frac{1}{2}} \times$$
$$\times \prod_{\substack{\ell=1\\ \ell\neq p}}^{N} \left|\frac{\zeta_p-\zeta_\ell^*}{\zeta_p-\zeta_\ell}\right| \prod_{\substack{k=1\\ k\neq j}}^{N} \left|\frac{\zeta_j-\zeta_k^*}{\zeta_j-\zeta_k}\right| \equiv M_{pj}.$$

For this purpose we develop some further notation.
By $\tilde{A}$ we denote the Gram matrix of the vectors $e_1, e_2, \ldots, e_{2N}$, introduced in (5.16a), i.e. $\tilde{A} = (\tilde{\alpha}_{rs})$, $\tilde{\alpha}_{rs} = \langle e_r, e_s\rangle$.
Since the vectors $e_1, e_2, \ldots, e_{2N}$ are linearly independent, it follows that $\det \tilde{A} > 0$ (see [3]).
Evidently

$$(5.22)\qquad \tilde{A} = \begin{pmatrix} Z^t & 0 \\ 0 & Z \end{pmatrix}, \qquad\qquad (\tilde{A})^{-1} = \begin{pmatrix} (Z^{-1})^t & 0 \\ 0 & Z^{-1} \end{pmatrix}.$$

Let us write $(\tilde{A})^{-1} = (\tilde{\beta}_{rs})$ and introduce the vectors $h_r = \sum_{s=1}^{2N} \tilde{\beta}_{rs} e_s$.
Note that $\langle h_r, e_s\rangle = \delta_{rs}$ and $\langle h_r, h_s\rangle = \tilde{\beta}_{rs}$.
In combination with (5.10-20) this gives

$$(5.23)\qquad (I - S)h_s = \sum_{r=1}^{2N} \beta_{rs} e_r.$$

Using the identity $I - S = T_d S$, we get

$$(5.24)\qquad \beta_{rs} = \langle T_d S h_s, h_r\rangle.$$

Hence, in view of (5.15)

$$(5.25)\qquad |\beta_{rs}|^2 \leq \|h_r\|_2^2 \|h_s\|_2^2 = \tilde{\beta}_{rr}\tilde{\beta}_{ss}, \qquad r,s = 1,2,\ldots,2N.$$

From (5.22) it is clear that for $\ell = 1,2,\ldots,N$

$$(5.26)\qquad \tilde{\beta}_{\ell\ell} = \tilde{\beta}_{\ell+N,\ell+N} = \frac{\det(\zeta_{pj})_{p,j=1,p\neq\ell,j\neq\ell}^{N}}{\det(\zeta_{pj})_{p,j=1}^{N}} = 4(\operatorname{Im}\zeta_\ell) \prod_{\substack{p=1\\ p\neq\ell}}^{N} \left|\frac{\zeta_\ell-\zeta_p^*}{\zeta_\ell-\zeta_p}\right|^2.$$

The desired estimate (5.21) is, of course, an immediate consequence of (5.25-26).

By (5.10b-21) we have

$$(5.27)\qquad \|A_p\|_E \leq \|f\| \sum_{j=1}^{N} (\sqrt{2}\ \mathrm{Im}\ \zeta_j)^{-1} M_{pj},$$

yielding the bound (5.12-13) for $\|S\|$. □

Corollary to lemma 5.1. *For each* $x \in \mathbb{R}$ *the equation*

$$(5.28)\qquad (I + T_d)\beta = -\omega_d$$

admits a unique solution $\beta_d \in \mathcal{B}$ *and we have*

$$(5.29)\qquad \beta_d(y;x) = -\sum_{p,j=1}^{N} \begin{pmatrix} \beta_{p,j+N} & \beta_{p+N,j+N} \\ \beta_{p,j} & \beta_{p+N,j} \end{pmatrix} \begin{pmatrix} 0 & \varepsilon_p^*(y) \\ \varepsilon_p(y) & 0 \end{pmatrix}$$

$$\equiv \begin{pmatrix} a_d^* & -b_d \\ b_d^* & a_d \end{pmatrix}.$$

Remark. Let us recall that β_d produces the reflectionless part of the potential q through the formula

$$(5.30)\qquad q_d(x) = b_d(0^+;x) = \sum_{p,j=1}^{N} \beta_{p,j+N}.$$

Clearly, by (5.21) we have the a priori bound

$$(5.31)\qquad \sup_{x\in\mathbb{R}} |q_d(x)| \leq \sum_{p,j=1}^{N} M_{pj},$$

which does not involve the C_j^r but depends only on the ζ_j in a simple explicit way.

6. Proof of theorem 4.1.

The nature of the results obtained in the previous section enables us to provide a proof of theorem 4.1 which is remarkably similar in form to that given in Chapter 4 for the corresponding scalar case.

Let $x \in \mathbb{R}$ be arbitrarily fixed.
To start with, let us write the solution β of (5.9) in the form

(6.1) $\beta = \beta_d + \beta_c$, with

(6.2) $\beta_d = -S\omega_d$.

By (3.4-5) and (5.29-30) we plainly have

(6.3) $$\beta_c = \begin{pmatrix} a_c^* & -b_c \\ b_c^* & a_c \end{pmatrix},$$

with $a_c = a - a_d$ and $b_c = b - b_d$, such that

(6.4a) $q(x) - q_d(x) = b_c(0^+;x)$

(6.4b) $$\int_x^\infty (|q(s)|^2 - |q_d(s)|^2)ds = -a_c(0^+;x).$$

From section 5 it is clear that both β and β_d belong to $\mathcal{B} \cap \mathcal{H}$. Hence, we already know that $\beta_c \in \mathcal{B} \cap \mathcal{H}$. It remains to find a concrete estimate of β_c in the norm (5.4). For that purpose we insert the decomposition (6.1) into (5.9), thereby obtaining

(6.5) $(I + T_c + T_d)\beta_c = -T_c\beta_d - \omega_c$.

Consider (6.5) as an equation in the Hilbert space $\mathcal{H}$. Since $T_c + T_d$ is antisymmetric, the operator $I + T_c + T_d$ is invertible on $\mathcal{H}$. Furthermore, the relation (5.14) holds with T_d replaced by $T_c + T_d$. Thus (6.5) has a unique solution $\beta_c \in \mathcal{H}$ satisfying

(6.6) $\|\beta_c\|_2 \leq \|T_c\beta_d\|_2 + \|\omega_c\|_2$.

Using the generalized Minkowski inequality (see [6], p. 148) we obtain

(6.7) $$\|T_c\beta_d\|_2 \leq \int_0^\infty \left(\int_0^\infty \|\beta_d(z;x)\omega_c(x+y+z)\|_E^2 dy\right)^{\frac{1}{2}} dz$$
$$\leq \int_0^\infty \|\beta_d(z;x)\|_E \left(\int_0^\infty \|\omega_c(x+y+z)\|_E^2 dy\right)^{\frac{1}{2}} dz.$$

Hence

(6.8) $\|T_c\beta_d\|_2 \leq \|\omega_c\|_2 \|\|\beta_d\|_1$, where by (5.21-29)

$$(6.9) \qquad \|\beta_d\|_1 \equiv \int_0^\infty \|\beta_d(z;x)\|_E dz \le \sum_{p,j=1}^{N} (\sqrt{2}\ \mathrm{Im}\ \zeta_p)^{-1} M_{pj}.$$

We conclude that

$$(6.10) \qquad \|\beta_c\|_2 \le \|\omega_c\|_2 (1 + \|\beta_d\|_1) \le \alpha_0 \|\omega_c\|_2$$

with α_0 given by (5.13a).
The trick is now to rewrite equation (6.5) as

$$(6.11) \qquad (I + T_d)\beta_c = -T_c\beta_c - T_c\beta_d - \omega_c$$

and to realize that the a priori estimate (6.10) paves the way to estimate the right hand side of (6.11) in the norm (5.4). In fact, since

$$(6.12) \qquad \|(T_c\beta_c)(y;x)\|_E \le \int_0^\infty \|\beta_c(z;x)\|_E \|\omega_c(x+y+z)\|_E dz,$$

we have by Schwarz' inequality

$$(6.13) \qquad \|T_c\beta_c\| \le \sup_{0<y<+\infty} \left(\int_0^\infty \|\beta_c(z;x)\|_E^2 dz\right)^{\frac{1}{2}} \left(\int_0^\infty \|\omega_c(x+y+z)\|_E^2 dz\right)^{\frac{1}{2}}$$

$$\le \|\beta_c\|_2 \|\omega_c\|_2 \le \|\omega_c\|_2^2 (1 + \|\beta_d\|_1).$$

Moreover, invoking again the generalized Minkowski inequality, one gets

$$(6.14) \qquad \|T_c\beta_d\| \le \|\omega_c\| \|\beta_d\|_1.$$

Together, (6.13) and (6.14) provide the estimate

$$(6.15) \qquad \|-T_c\beta_c - T_c\beta_d - \omega_c\| \le (\|\omega_c\| + \|\omega_c\|_2^2)(1 + \|\beta_d\|_1).$$

Applying lemma 5.1 we obtain from (6.11-15) the following estimate for β_c in the norm (5.4)

$$(6.16) \qquad \|\beta_c\| \le \alpha_0^2 (\|\omega_c\| + \|\omega_c\|_2^2).$$

Insertion of (5.6c) and (6.3) then leads to

$$(6.17) \qquad \sup_{0<y<+\infty} \left(|a_c(y;x)|^2 + |b_c(y;x)|^2\right)^{\frac{1}{2}} \le \alpha_0^2 \Big(\sqrt{2} \int_0^\infty |\Omega_c(x+y)|^2 dy +$$

$$+ \sup_{0<y<+\infty} |\Omega_c(x+y)|\Big).$$

Combining (6.17) with (6.4a) we arrive at the desired estimate (4.1-2), wherewith the proof of theorem 4.1 is completed.

Remark. Actually, we have proven more, since by (6.4b) and (6.17) the estimate (4.12) still holds if one replaces the left hand side of (4.1) by

$$(6.18) \qquad \max\left[|q(x) - q_d(x)|, \left| \int_x^\infty \left(|q(s)|^2 - |q_d(s)|^2 \right) ds \right| \right].$$

7. An application: cmKdV asymptotics.

The importance of theorem 4.1 stems from the fact that it is a useful tool to investigate the asymptotic behaviour of solutions of certain non-linear evolution equations solvable by the Zakharov-Shabat inverse scattering method (see the discussion in Chapter 4, section 4).

As an illustration let us consider the complex modified Korteweg-de Vries (cmKdV) problem

$$(7.1a) \qquad q_t + 6|q|^2 q_x + q_{xxx} = 0, \qquad -\infty < x < +\infty, \qquad t > 0$$

$$(7.1b) \qquad q(x,0) = q_0(x),$$

where the initial function $q_0(x)$ is an arbitrary complex-valued function on $\mathbb{R}$, such that

(7.2a) $q_0(x)$ satisfies the hypotheses (2.2-11) and is therefore a bona fide potential in the Zakharov-Shabat scattering problem (2.1).

(7.2b) $q_0(x)$ is sufficiently smooth and (along with a number of its derivatives) decays sufficiently rapidly for $|x| \to \infty$:

(i) for the whole of the Zakharov-Shabat inverse scattering method [2] to work,

(ii) to guarantee certain regularity and decay properties of the right reflection coefficient to be stated further on.

Uniqueness of solutions of (7.1) can be established within the class of functions which, together with a sufficient number of derivatives vanish for $|x| \to \infty$ (cf. [7]).

Suitably adapting the procedure outlined in [10] for the real mKdV problem one can prove by an inverse scattering analysis that condition (7.2) guarantees the existence of a complex-valued function $q(x,t)$, continuous on $\mathbb{R}\times[0,\infty)$, such that

(7.3a) For any value of the time $t \geq 0$ the function $q(x,t)$ satisfies the hypotheses (2.2-11).

(7.3b) $q(x,t)$ satisfies (7.1) in the classical sense.

(7.3c) $q(x,t)$ falls in the class of functions for which uniqueness of solutions of (7.1) can be proven.

Whenever, in the sequel, we speak of "the solution" of (7.1) we refer to the solution obtained by inverse scattering.

Uniqueness implies that if $q_0(x)$ is a real-valued function then so is $q(x,t)$ for all $t > 0$. Thus, in that case, the cmKdV problem (7.1) reduces to the mKdV problem (1.1) in [9], i.e. Chapter 5.

Let us point out that by the inverse scattering method the solution $q(x,t)$ of (7.1) is obtained as follows.
Having computed the (right) scattering data $\{b_r(\zeta),\zeta_j,C_j^r\}$ associated with $q_0(x)$, one puts (see [4], p. 307)

(7.4a) $$C_j^r(t) = C_j^r\exp\{8i\zeta_j^3t\}, \qquad j = 1,2,\dots,N$$

(7.4b) $$b_r(\zeta,t) = b_r(\zeta)\exp\{8i\zeta^3t\}, \qquad -\infty < \zeta < +\infty.$$

Then by the solvability of the inverse scattering problem, there exists for each $t > 0$ a smooth potential $q(x,t)$ satisfying the hypotheses (2.2-11) and having $\{b_r(\zeta,t),\zeta_j,C_j^r(t)\}$ as its scattering data. The function $q(x,t)$ is the unique solution of the cmKdV initial value problem (7.1).

Plainly, the reflectionless part $q_d(x,t)$ of $q(x,t)$ can be obtained in explicit form by substituting (7.4a) into (5.30).

In the N = 1 case, setting $\zeta_1 = \xi + i\eta$, $C_1^r = \rho e^{i\phi}$, with $\eta, \rho > 0$, $\xi, \phi \in \mathbb{R}$, this yields

(7.5) $$q_d(x,t) = 2\eta e^{-i\Phi}\operatorname{sech}\Psi,$$

with

(7.6a) $$\Phi = 2\xi x + 8\xi(\xi^2 - 3\eta^2)t + \phi + \frac{\pi}{2}$$

(7.6b) $$\Psi = 2\eta x + 8\eta(3\xi^2 - \eta^2)t + \psi$$

(7.6c) $$\psi = \log\{\frac{2\eta}{\rho}\}.$$

Thus the one-soliton solution is a single wave packet modulated by an envelope having the shape of a hyperbolic secant. The envelope and phase velocities are found from (7.6) to be $v_e = 4(\eta^2 - 3\xi^2)$ and $v_{ph} = 4(3\eta^2 - \xi^2)$, respectively. According to the sign of v_e the envelope may propagate to the right, to the left or be at rest.

If N > 1, then, generically, for large time $q_d(x,t)$ will decompose into N distinct solitons of the structure (7.5). However, it is easy to construct examples (e.g. N = 2, $\zeta_1 = \xi + i\eta$, $\eta^2 > 3\xi^2$, $\zeta_2 = i(\eta^2 - 3\xi^2)^{\frac{1}{2}}$) in which no such decomposition takes place, but a more complicated structure is found instead.

Because of the negative group velocity associated with the linearized version of (7.1) we expect that for large t, when considered on the positive x-axis, the solution q(x,t) of (7.1) is approximated with reasonable accuracy by its reflectionless part $q_d(x,t)$.
We shall use theorem 4.1 to verify this.

Note first that, if $q_0(x)$ satisfies (7.2a-b(i)), then by (7.3a) the function q(x,t) satisfies the conditions of theorem 4.1.
Consequently, for each $x \in \mathbb{R}$ and $t > 0$ one has

(7.7) $$|q(x,t) - q_d(x,t)| \leq \alpha_0^2\Big(\sqrt{2}\int_0^\infty |\Omega_c(x+y;t)|^2 dy + \\ + \sup_{0<y<+\infty} |\Omega_c(x+y;t)|\Big),$$

with

$$(7.8) \qquad \Omega_c(s;t) = \frac{1}{\pi} \int_{-\infty}^{\infty} b_r(\zeta) e^{2i\zeta s + 8i\zeta^3 t} d\zeta, \qquad s \in \mathbb{R}$$

and α_0 the constant given by (4.2).

Next, suppose that $q_0(x)$ satisfies (7.2) in such a way that

(7.9) b_r is of class $C^2(\mathbb{R})$ and the derivatives $b_r^{(j)}(\zeta)$, $j = 0,1,2$ are bounded on $\mathbb{R}$.

Then, reasoning as in Chapter 2, but taking into account that the symmetry relation $b_r^*(\zeta) = b_r(-\zeta)$ is no longer guaranteed, it is easily seen that in the parameter region

(7.10) $T = (3t)^{1/3} \geq 1$, $x \geq -\mu - \nu T$, where μ and ν are nonnegative constants

one has the estimates

$$(7.11a) \qquad \sup_{0<y<+\infty} |\Omega_c(x+y;t)| \leq \gamma T^{-1}$$

$$(7.11b) \qquad \int_0^{\infty} |\Omega_c(x+y;t)| dy \leq |b_r(0)| \left(\frac{1}{3} + \int_{-\nu}^{0} |Ai(\eta)| d\eta \right) + \gamma T^{-1}$$

where γ is some constant.

Combining (7.7) with the estimates (7.11) we arrive at

Theorem 7.1. *Let $q(x,t)$ be the solution of the complex modified Korteweg-de Vries problem*

$$(7.12) \qquad \begin{cases} q_t + 6|q|^2 q_x + q_{xxx} = 0, & -\infty < x < +\infty, \quad t > 0 \\ q(x,0) = q_0(x), \end{cases}$$

where the initial function $q_0(x)$ is an arbitrary complex-valued function on $\mathbb{R}$, satisfying (7.2) in such a way that (7.9) is fulfilled. Let $\{b_r(\zeta), \zeta_j, c_j^r\}$ be the scattering data associated with $q_0(x)$. Then for each $x \in \mathbb{R}$ and $t > 0$ one has

$$(7.13) \qquad |q(x,t) - q_d(x,t)| \leq \alpha_0^2 \Big(\sqrt{2} \int_0^{\infty} |\Omega_c(x+y;t)|^2 dy + \\ + \sup_{0<y<+\infty} |\Omega_c(x+y;t)| \Big),$$

with $q_d(x,t)$ *the reflectionless part (5.30), (7.4a) of* $q(x,t)$, α_0 *the constant given by (4.2) and* Ω_c *the function introduced in (7.8).*

Next, let μ *and* ν *be arbitrary nonnegative constants. Put* $\alpha = \mu + \nu T$, $T = (3t)^{1/3}$. *Then the following estimate holds*

(7.14a) $$\sup_{x \geq -\alpha} |q(x,t) - q_d(x,t)| \leq A, \qquad \textit{for } t > 0$$

(7.14b) $$\sup_{x \geq -\alpha} |q(x,t) - q_d(x,t)| \leq \tilde{\gamma} T^{-1}, \qquad \textit{for } t \geq \frac{1}{3}$$

with

(7.15a) $$A = \frac{\alpha_0^2}{\pi} \int_0^{\infty} \left(|b_r(\zeta)| + \sqrt{2} |b_r(\zeta)|^2 \right) d\zeta$$

(7.15b) $$\tilde{\gamma} = \alpha_0^2 \gamma \left(1 + \sqrt{2}\gamma + \sqrt{2} |b_r(0)| \left(\frac{1}{3} + \int_{-\nu}^{0} |Ai(\eta)| d\eta \right) \right)$$

where γ *denotes the constant appearing in (7.11).*

Clearly, theorem 7.1 generalizes theorem 5.1 in Chapter 5 obtained for the real case. Generalizations of other results can be derived in a similar way, but they fall outside the scope of these notes and are therefore left to the interested reader.

References

[1] M.J. Ablowitz, Lectures on the inverse scattering transform, Stud. Appl. Math. 58 (1978), 17-94.

[2] M.J. Ablowitz, D.J. Kaup, A.C. Newell and H. Segur, The inverse scattering transform - Fourier analysis for nonlinear problems, Stud. Appl. Math. 53 (1974), 249-315.

[3] P.J. Davis, Interpolation and Approximation, Dover, New York, 1963.

[4] R.K. Dodd, J.C. Eilbeck, J.D. Gibbon and H.C. Morris, Solitons and Nonlinear Wave Equations, Academic Press, 1982.

[5] W. Eckhaus and A. van Harten, The Inverse Scattering Transformation and the Theory of Solitons, North-Holland Mathematics Studies 50, 1981 (2nd ed. 1983).

[6] G.H. Hardy, J.E. Littlewood and G. Pólya, Inequalities, 2nd ed., Cambridge 1952.

[7] Y. Kametaka, Korteweg-de Vries equation IV. Simplest generalization, Proc. Japan Acad., 45 (1969), 661-665.

[8] P. Schuur, On the approximation of a real potential in the Zakharov-Shabat system by its reflectionless part, preprint 341, Mathematical Institute Utrecht (1984).

[9] P. Schuur, Decomposition and estimates of solutions of the modified Korteweg-de Vries equation on right half lines slowly moving leftward, preprint 342, Mathematical Institute Utrecht (1984).

[10] S. Tanaka, Non-linear Schrödinger equation and modified Korteweg-de Vries equation; construction of solutions in terms of scattering data, Publ. R.I.M.S. Kyoto Univ. 10 (1975), 329-357.

[11] V.E. Zakharov and A.B. Shabat, Exact theory of two-dimensional self-focusing and one-dimensional self-modulation of waves in non-linear media, Soviet Phys. JETP (1972), 62-69.

CONCLUDING REMARKS

In what follows we have collected a number of additional results, each with some interest of its own. We present these results without proof.

(i) Let $u(x,t)$ be a real solution of the KdV obtained via IST as described in Chapter 2. Suppose there are $x_1, x_2 \in \mathbb{R}$, $x_1 \neq x_2$ and $t_2 > t_1 > 0$, such that for all $x \in \mathbb{R}$

$$u(x_i+x, t_i) = u(x_i-x, t_i), \quad i = 1,2.$$

Then one has:

(a) If $x_1 > x_2$, then $u \equiv 0$.

(b) If $x_1 < x_2$, then either $u \equiv 0$ or $u(x,t) = -2\kappa^2 \operatorname{sech}^2[\kappa(x-x^+-4\kappa^2 t)]$ with

$$\kappa = \tfrac{1}{2}\sqrt{\frac{x_2-x_1}{t_2-t_1}} \quad \text{and} \quad x^+ = \frac{x_1 t_2 - x_2 t_1}{t_2-t_1}.$$

In particular this shows that, apart from the 1-soliton solution, any nontrivial solution of the KdV obtained via IST can have spatial symmetry for at most one value of the time t.

(ii) By combining the remark made in Chapter 2 after the proof of lemma 5.1 with the other results from that chapter one easily arrives at the following theorem:

Consider a t*-parameter family* $u(x,t)$, $t \geq t_0$, $t_0 \in \mathbb{R}$ *of real potentials in the Schrödinger scattering problem, satisfying for fixed* $t \geq t_0$ *the conditions stated in Ch.* 2, *subsection* 2.1. *Let* $\{b_r(k,t), \kappa_j(t), c_j^r(t)\}$ *be the scattering data associated with* $u(x,t)$. *Write* $u_d(x,t)$ *for the reflectionless part of* $u(x,t)$, *i.e. the potential with scattering data* $\{0, \kappa_j(t), c_j^r(t)\}$. *Let* Ω_c *be given by Ch.* 2, (2.26c) *and let* V *be as on p.* 45.

Assume there are constants $c_0, c_1, \ldots, c_N$, *such that*

$$0 < c_N < \kappa_N(t) < c_{N-1} < \ldots < c_2 < \kappa_2(t) < c_1 < \kappa_1(t) < c_0.$$

Assume furthermore, that the function $y \mapsto \Omega_c(x+y;t)$ *is strongly differentiable in* V *with respect to* x *at every point* (x,t), $x \in \mathbb{R}$, $t \geq t_0$. *Let there exist a function* $\alpha:[t_0,\infty) \to \mathbb{R}$, *such that in the parameter region* $t \geq t_0$, $x \geq \alpha(t)$, *the functions* Ω_c *and* Ω_c', *with* Ω_c' *the strong* x*-derivative of* Ω_c, *satisfy:*

(a) $\max\left[|\Omega_c(x+y;t)|, |\Omega_c'(x+y;t)|\right] \leq H(y,t), \quad y > 0,$
with H(y,t) *a monotonically decreasing function of* y *for fixed* t, *such that* $\sigma(t) \equiv \sup_{0<y<+\infty} H(y,t) \to 0$ *as* $t \to \infty$.

(b) $\int_0^\infty |\Omega_c(x+y;t)|dy \leq \sigma_0 < 1$ *and*
$\int_0^\infty |\Omega_c'(x+y;t)|dy \leq \sigma_1 < +\infty$, *with* σ_0 *and* σ_1 *constants.*

Then we have:

$$\sup_{x \geq \alpha(t)} |u(x,t)-u_d(x,t)| = O(\sigma(t)) \qquad \text{as } t \to \infty.$$

(iii) The inverse scattering formalism associated with the self-adjoint Zakharov-Shabat system

$$\begin{pmatrix}\psi_1\\ \psi_2\end{pmatrix}' = \begin{pmatrix}-i\zeta & q\\ q & i\zeta\end{pmatrix}\begin{pmatrix}\psi_1\\ \psi_2\end{pmatrix}, \qquad ' = \frac{d}{dx}, \qquad -\infty < x < +\infty,$$

where $q = q(x)$ is a real function and ζ a complex parameter, can be simplified. As in Chapter 4 the Gel'fand-Levitan equation appearing in the literature can be reduced to a scalar integral equation containing only a single integral. With the help of this simplification one can analyse in a way similar to Ch. 2, section 4, the asymptotic behaviour corresponding with the various nonlinear evolution equations (all solitonless) solvable via the associated inverse scattering method, such as the solitonless mKdV, sinh-Gordon, etc.

(iv) The explicit structure of the constants in the mKdV analysis presented in Chapter 5, shows that related results are valid in coordinate regions $t > 0$, $x \geq -\mu - \nu t^{\delta}$, $\delta = \frac{1}{3} + \varepsilon$, with μ,ν,ε nonnegative constants and $\varepsilon > 0$ sufficiently small.

APPENDIX

AN OPEN PROBLEM

In this volume we have succeeded, by a more or less uniform method, to reveal the asymptotic structure for large time for solutions of a number of interesting nonlinear evolution equations solvable by the inverse scattering method. However, any reader familiar with the soliton field no doubt has noticed the absence of the wellknown nonlinear Schrödinger equation (NLS)

(1a) $$iq_t = q_{xx} + 2q^2 q^*, \quad -\infty < x < +\infty, \quad t > 0$$

(1b) $$q(x,0) = q_0(x),$$

where the initial function $q_0(x)$ is an arbitrary complex-valued function on $\mathbb{R}$, sufficiently smooth and rapidly decaying for $|x| \to \infty$ and satisfying conditions similar to Ch. 8, (7.2) so as to make the Zakharov-Shabat inverse scattering method associated with Ch. 8, (2.1) work.

Note that the dispersion relation $\omega(\zeta) = -\zeta^2$ associated with the linearized version of (1a) is an even function, so that the group velocity $d\omega/d\zeta = -2\zeta$ is not of one sign for all real ζ. This forms a significant difference with the other problems treated and has direct consequences for the method employed in this volume.

Let us nevertheless see what our method produces.

To start with let us point out that for the problem (1) the Zakharov-Shabat inverse scattering method runs as follows (see [1], [2], [4], [6] for details).

Having computed the (right) scattering data $\{b_r(\zeta),\zeta_j,C_j^r\}$ associated with $q_0(x)$, when introduced as a potential into Ch. 8, (2.1), one puts

$$C_j^r(t) = C_j^r\exp\{-4i\zeta_j^2t\}, \qquad j = 1,2,\ldots,N \tag{2a}$$

$$b_r(\zeta,t) = b_r(\zeta)\exp\{-4i\zeta^2t\}, \qquad -\infty < \zeta < +\infty. \tag{2b}$$

Then by the solvability of the inverse scattering problem, there exists for each $t > 0$ a smooth potential $q(x,t)$ having $\{b_r(\zeta,t),\zeta_j,C_j^r(t)\}$ as its scattering data. The function $q(x,t)$ is the unique solution to the NLS initial value problem (1).

Now, let $q_d(x,t)$ denote the reflectionless part of the solution $q(x,t)$ of (1). Then, since the group velocity is not of constant sign, there is no particular region of the x-axis singled out on which $q(x,t) - q_d(x,t)$ might be concentrated as $t \to \infty$. However, we may expect an overall decay of $q(x,t) - q_d(x,t)$ for $t \to \infty$. In fact, some authors [3], [5] claim the existence of special solutions to (1) such that $q - q_d$ decays in time as $t^{-\frac{1}{2}}$ more or less uniformly in x on $\mathbb{R}$.

Let us see what the techniques of Chapter 8 yield in this case. By theorem 4.1 of that chapter we can estimate $q - q_d$ in terms of the scattering data in the following way:

For each $x \in \mathbb{R}$ and $t > 0$ one has

$$|q(x,t) - q_d(x,t)| \leq \alpha_0^2\left(\sqrt{2}\int_0^\infty |\Omega_c(x+y;t)|^2\,dy + \sup_{0<y<+\infty} |\Omega_c(x+y;t)|\right), \tag{3}$$

with

$$\Omega_c(s;t) = \frac{1}{\pi}\int_{-\infty}^{\infty} b_r(\zeta)e^{2i\zeta s-4i\zeta^2t}\,d\zeta, \qquad s \in \mathbb{R} \tag{4}$$

and α_0 the constant given by Ch. 8, (4.2).

Now, a detailed examination of the right hand side of (3) shows that in all relevant coordinate regions the second term behaves as $t^{-\frac{1}{2}}$ as $t \to \infty$, but the first term tends to a constant!

Note the difference with the cmKdV problem Ch. 8, (7.1) where theorem 4.1

revealed neatly the asymptotic structure (see Ch. 8, (7.7-8) and the subsequent discussion).

As an explicit prototype problem, let us mention the special case that the initial function $q_0(x)$ in (1b) has the scattering data $\{b_r(\zeta) = e^{-\zeta^2}, 0, 0\}$. Then the integral (4) can be evaluated in closed form, yielding

$$(5) \qquad \Omega_c(s;t) = \pi^{-\frac{1}{2}}(1 + 4it)^{-\frac{1}{2}}\exp\{-s^2/(1+4it)\}.$$

Let us write out in full glory the Gel'fand-Levitan equation Ch. 8, (3.2-3) in the special case (5). For $x \in \mathbb{R}$, $y > 0$ and $t > 0$ one then has

$$(6a) \qquad \beta(y;x,t) + \omega(x+y;t) + \int_0^\infty \beta(z;x,t)\omega(x+y+z;t)dz = 0$$

$$(6b) \qquad \omega(s;t) = \begin{pmatrix} 0 & -\pi^{-\frac{1}{2}}(1-4it)^{-\frac{1}{2}}\exp\{-s^2/(1-4it)\} \\ \pi^{-\frac{1}{2}}(1+4it)^{-\frac{1}{2}}\exp\{-s^2/(1+4it)\} & 0 \end{pmatrix}$$

with $s \in \mathbb{R}$.

Here the unknown $\beta(y;x,t)$ is a 2×2 matrix function of the variable y, whereas x and t are parameters. Of crucial importance is the behaviour of $\beta(0^+;x,t)$ as $t \to +\infty$ in appropriate regions $x \geq \alpha(t)$. This gives the behaviour of $q(x,t)$ through the relations Ch. 8, (3.4-5).

Since the integral equation (6) is an explicit integral equation, no particular knowledge of inverse scattering is required to attack it.

References

[1] M.J. Ablowitz, Lectures on the inverse scattering transform, Stud. Appl. Math. 58 (1978), 17-94.

[2] M.J. Ablowitz, D.J. Kaup, A.C. Newell and H. Segur, The inverse scattering transform - Fourier analysis for nonlinear problems, Stud. Appl. Math. 53 (1974), 249-315.

[3] M.J. Ablowitz and H. Segur, Asymptotic solutions and conservation laws for the nonlinear Schrödinger equation I, J. Math. Phys., 17 (1976), 710-713.

[4] W. Eckhaus and A. van Harten, The Inverse Scattering Transformation and the Theory of Solitons, North-Holland Mathematics Studies 50, 1981 (2nd ed. 1983).

[5] H. Segur, Asymptotic solutions and conservation laws for the nonlinear Schrödinger equation II, J. Math. Phys., 17 (1976), 714-716.

[6] S. Tanaka, Non-linear Schrödinger equation and modified Korteweg-de Vries equation, construction of solutions in terms of scattering data, Publ. R.I.M.S. Kyoto Univ. 10 (1975), 329-357.

INDEX